ENVIRONMENTAL SCIENCE, ENGINEERING AND TECHNOLOGY

CONTAMINATED DEPARTMENT OF DEFENSE SITE: BACKGROUND AND ISSUES

ENVIRONMENTAL SCIENCE, ENGINEERING AND TECHNOLOGY

Additional books in this series can be found on Nova's website under the Series tab.

Additional E-books in this series can be found on Nova's website under the E-books tab.

Environmental Science, Engineering and Technology

Contaminated Department of Defense Site: Background and Issues

Douglas B. Ferro
Editor

Nova Science Publishers, Inc.
New York

Copyright © 2011 by Nova Science Publishers, Inc.

All rights reserved. No part of this book may be reproduced, stored in a retrieval system or transmitted in any form or by any means: electronic, electrostatic, magnetic, tape, mechanical photocopying, recording or otherwise without the written permission of the Publisher.

For permission to use material from this book please contact us:
Telephone 631-231-7269; Fax 631-231-8175
Web Site: http://www.novapublishers.com

NOTICE TO THE READER

The Publisher has taken reasonable care in the preparation of this book, but makes no expressed or implied warranty of any kind and assumes no responsibility for any errors or omissions. No liability is assumed for incidental or consequential damages in connection with or arising out of information contained in this book. The Publisher shall not be liable for any special, consequential, or exemplary damages resulting, in whole or in part, from the readers' use of, or reliance upon, this material. Any parts of this book based on government reports are so indicated and copyright is claimed for those parts to the extent applicable to compilations of such works.

Independent verification should be sought for any data, advice or recommendations contained in this book. In addition, no responsibility is assumed by the publisher for any injury and/or damage to persons or property arising from any methods, products, instructions, ideas or otherwise contained in this publication.

This publication is designed to provide accurate and authoritative information with regard to the subject matter covered herein. It is sold with the clear understanding that the Publisher is not engaged in rendering legal or any other professional services. If legal or any other expert assistance is required, the services of a competent person should be sought. FROM A DECLARATION OF PARTICIPANTS JOINTLY ADOPTED BY A COMMITTEE OF THE AMERICAN BAR ASSOCIATION AND A COMMITTEE OF PUBLISHERS.

Additional color graphics may be available in the e-book version of this book.

LIBRARY OF CONGRESSCATALOGING-IN-PUBLICATION DATA
Contaminated Department of Defense site : background and issues / editor,
Douglas B. Ferro.
p. cm.
Includes index.
ISBN 978-1-61122-465-8 (hardcover)
1. Hazardous waste site remediation--United States. 2. Hazardous waste
sites--United States. 3. United States. Dept. of Defense--Buildings. I.
Ferro, Douglas B. II. United States. Dept. of Defense.
TD1040.C66 2010
363.72'870973--dc22
2010037353

Published by Nova Science Publishers, Inc. † New York

CONTENTS

Preface   **vii**

Chapter 1   Superfund: Greater EPA Enforcement and
              Reporting are Needed to Enhance Cleanup at DOD Sites   **1**
              United States Government Accountability Office

Chapter 2   Superfund: Interagency Agreements and
              Improved Project Management Needed
              to Achieve Cleanup Progress at Key
              Defense Installations   **45**
              United States Government Accountability Office

Index   **113**

PREFACE

Prior to the 1980s and the passage of environmental legislation regulating the generation, storage, and disposal of hazardous waste, Department of Defense (DOD) activities and industrial facilities contaminated millions of acres of soil and water on and near DOD sites in the United States and its territories. DOD installations generate hazardous wastes primarily through industrial operations to repair and maintain military equipment. Manufacturing and testing weapons at Army ammunition plants and proving grounds have caused some serious contamination problems as well. This book examines the background and issues of contaminated Department of Defense sites across the United States and the issues for the EPA in enforcing and reporting these sites to achieve cleanup progress at these key defense installations.

Chapter 1- Prior to the 1980s and the passage of environmental legislation—particularly the Comprehensive Environmental Response, Compensation, and Liability Act (CERCLA) governing environmental cleanup—Department of Defense (DOD) activities contaminated millions of acres of soil and water on and near DOD sites. The Environmental Protection Agency (EPA), which enforces CERCLA, places the most contaminated sites on its National Priorities List (NPL) and requires that they be cleaned up in accordance with CERCLA. EPA has placed 140 DOD sites on the NPL. Disputes have recently arisen between EPA and DOD on agreements to clean up some of these sites. In addition, most sites were placed on the NPL before 1991; since fiscal year 2000, EPA has added five DOD sites. In this context, we agreed to determine (1) the extent of EPA's oversight during assessment and cleanup at DOD sites and (2) why EPA has proposed fewer DOD sites for the NPL since the early 1990s. GAO interviewed officials at EPA and DOD and reviewed site file documentation at four EPA regions.

Chapter 2- Before the passage of federal environmental legislation in the 1970s and 1980s, Department of Defense (DOD) activities contaminated millions of acres of soil and water on and near DOD sites. The Environmental Protection Agency (EPA) has certain oversight authorities for cleaning up contaminants on federal property, and has placed 1,620 of the most contaminated sites—including 141 DOD installations—on its National Priorities List (NPL). As of February 2009, after 10 or more years on the NPL, 11 DOD installations had not signed the required interagency agreements (IAG) to guide cleanup with EPA. GAO was asked to examine (1) the status of DOD cleanup of hazardous substances at selected installations that lacked IAGs, and (2) obstacles, if any, to cleanup at these installations. GAO selected and visited three installations, reviewed relevant statutes and agency documents, and interviewed agency officials.

In: Contaminated Department of Defense Site...   ISBN: 978-1-61122-465-8
Editors: Douglas B. Ferro   © 2011 Nova Science Publishers, Inc.

Chapter 1

SUPERFUND: GREATER EPA ENFORCEMENT AND REPORTING ARE NEEDED TO ENHANCE CLEANUP AT DOD SITES*

United States Government Accountability Office

WHY GAO DID THIS STUDY

Prior to the 1980s and the passage of environmental legislation—particularly the Comprehensive Environmental Response, Compensation, and Liability Act (CERCLA) governing environmental cleanup—Department of Defense (DOD) activities contaminated millions of acres of soil and water on and near DOD sites. The Environmental Protection Agency (EPA), which enforces CERCLA, places the most contaminated sites on its National Priorities List (NPL) and requires that they be cleaned up in accordance with CERCLA. EPA has placed 140 DOD sites on the NPL. Disputes have recently arisen between EPA and DOD on agreements to clean up some of these sites. In addition, most sites were placed on the NPL before 1991; since fiscal year 2000, EPA has added five DOD sites. In this context, we agreed to determine (1) the extent of EPA's oversight during assessment and cleanup at DOD sites and (2) why EPA has proposed fewer DOD sites for the NPL since the early

---

*This is an edited, reformatted and augmented edition of a United States Government Accountability Office publication, Report GAO-09-278, dated March 2009.

2   United States Government Accountability Office

1990s. GAO interviewed officials at EPA and DOD and reviewed site file documentation at four EPA regions.

WHAT GAO RECOMMENDS

GAO suggests that Congress consider amending CERCLA to expand EPA's enforcement authority. EPA agreed that such authority would help assure timely and protective cleanup. DOD disagreed, stating that EPA has sufficient involvement. We continue to assert that EPA needs additional authority to ensure that cleanups are being done properly.

WHAT GAO FOUND

EPA evaluates DOD's preliminary assessments of contaminated DOD sites but has little to no oversight of the cleanup of the majority of these sites because most are not on the NPL. Of the 985 DOD sites requiring cleanup of hazardous substances, EPA has oversight authority of the 140 on the NPL; the remaining 845 non-NPL sites are overseen by other cleanup authorities— usually the states. Our review of 389 non-NPL DOD sites showed that EPA decided not to list 56 percent because it determined the condition of the sites did not satisfy the criteria for listing or because it deferred the sites to other programs, most often the Resource Conservation and Recovery Act—another federal statute that governs activities involving hazardous waste. However, EPA regional officials were unable to provide a rationale for not listing the remaining 44 percent because site files documenting EPA's decisions were missing or inconclusive. In addition, EPA has agreements with DOD for cleaning up 129 of the 140 NPL sites and is generally satisfied with the cleanup of these sites. However, DOD does not have agreements for the remaining 11 sites, even though they are required under CERCLA. It was not until more than 10 years after these sites were placed on the NPL that EPA, in 2007, pursued enforcement action against DOD by issuing administrative orders at 4 of the 11 sites.

Since the mid-1990s, EPA has placed fewer DOD sites on the NPL than in previous years for three key reasons. First, EPA does not generally list DOD sites that are being addressed under other federal or state programs to avoid duplication. Second, DOD and EPA officials told us that, because DOD has been identifying and cleaning up hazardous releases for more than two decades, and improved its management of waste generated during its ongoing operations, DOD has discovered fewer hazardous substance releases in recent years, making fewer sites available for listing. Third, in a few instances, state officials or others have objected to EPA's proposal to list contaminated DOD sites, and EPA has usually declined to proceed further. For example, in five instances EPA proposed contaminated DOD sites for the NPL that were not ultimately placed on the list. At four of these sites, the states' governors did not support listing, citing the perceived stigma of inclusion on the NPL and potential adverse economic effect. EPA did not list the fifth site because, according to EPA regional officials, DOD objected and appealed to the Office of Management and Budget, which recommended deferring this listing for 6 months to give DOD time to address personnel and contractor changes and demonstrate remediation progress. EPA officials recently told us that cleanup has taken place at these sites and that it was unlikely or unclear whether they would qualify for placement on the NPL based on their current condition.

ABBREVIATIONS

CERCLA	Comprehensive Environmental Response, Compensation, and Liability Act
DOD	Department of Defense
DOJ	Department of Justice
EPA	Environmental Protection Agency
HRS	Hazard Ranking System
IAG	interagency agreement
NCP	National Oil and Hazardous Substances Pollution Contingency Plan (National Contingency Plan)
NPL	National Priorities List
OMB	Office of Management and Budget
PCB	polychlorinated biphenyl
RCRA	Resource Conservation and Recovery Act
SARA	Superfund Amendments and Reauthorization Act
TCE	trichloroethylene

United States Government Accountability Office

March 13, 2009

The Honorable Edward J. Markey, Chairman
Subcommittee on Energy and Environment
Committee on Energy and Commerce
House of Representatives

The Honorable John D. Dingell
The Honorable Gene Green
House of Representatives

Prior to the 1980s and the passage of environmental legislation regulating the generation, storage, and disposal of hazardous waste, Department of Defense (DOD) activities and industrial facilities contaminated millions of acres of soil and water on and near DOD sites in the United States and its territories. DOD installations generate hazardous wastes primarily through industrial operations to repair and maintain military equipment. Manufacturing and testing weapons at Army ammunition plants and proving grounds have caused some serious contamination problems as well. To address the cleanup of hazardous substance releases nationwide, in 1980, Congress passed the Comprehensive Environmental Response, Compensation, and Liability Act (CERCLA), better known as "Superfund."

In 1986, CERCLA was amended by the Superfund Amendments and Reauthorization Act (SARA). SARA reflected concern with the adequacy and timeliness of DOD and other federal agency cleanups, which was compounded by the Environmental Protection Agency's (EPA) unwillingness or inability to carry out enforcement actions against other federal agencies. SARA addresses this problem by (1) requiring DOD and other federal agencies to comply with CERCLA; (2) providing EPA with the authority to select remedies at federal facility National Priorities List (NPL) sites if agreement cannot be reached on the remedy to be selected; and (3) requiring federal agencies to enter into interagency agreements (IAG) with EPA at NPL sites. SARA also added a citizen suit provision to CERCLA specifically authorizing nonfederal parties such as states and citizens' groups to sue DOD and other federal agencies to enforce the terms of IAGs, among other things; and established a Defense Environmental Restoration Program along with separate Department of the Treasury accounts specifically for DOD environmental cleanup activities—to better ensure cleanup funding availability—and requiring DOD to carry out those activities in accordance with CERCLA.

Section 120 of CERCLA, as amended, requires federal agencies to comply with CERCLA and submit information to EPA on certain potentially hazardous releases. EPA maintains this information in a Federal Agency Hazardous Waste Compliance Docket which includes a history of federal facilities that generate, transport, store, or dispose of hazardous waste or that have had some type of hazardous substance release or spill.

For each site on the docket, CERCLA Section 120 requires EPA to take steps to ensure that a preliminary site assessment is conducted by the responsible federal agency.[1] The preliminary assessment, which is generally based on site records and other information regarding hazardous substances stored or disposed of at the facility, forms the basis for EPA to evaluate the site for listing on the NPL. EPA reviews preliminary site assessments to determine whether a site poses little or no threat to human health and the environment or requires further investigation or assessment for possible cleanup. Based on this assessment, EPA may then score and rank the site based on whether the contamination presents a potential threat to human health and the environment.[2] If a site scores at or above a minimum threshold for cleanup under CERCLA, EPA may place the site on the NPL or defer it to another regulatory authority, such as a state agency, for cleanup under other statutory authorities or programs, such as the Resource Conservation and Recovery Act (RCRA). As of November 2008, the NPL included 1,587 sites.[3] Of these, according to EPA officials, 140 were federal DOD sites, representing almost 9 percent of the NPL.[4]

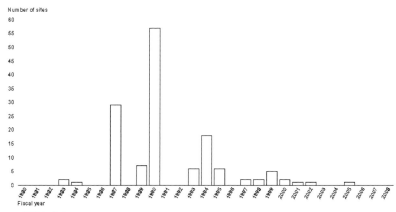

Source: EPA data.

Figure 1. Annual Number of DOD Sites Placed on the NPL Has Declined Since the 1990s, Fiscal Years 1980–2008

6 United States Government Accountability Office

Section 120 of CERCLA also establishes specific procedures for cleaning up federal facilities on the NPL. As part of its oversight responsibility, EPA works with DOD to evaluate the nature and extent of contamination at a site, select a remedy, track cleanup and monitor the remedy's effectiveness in protecting human health and the environment. Under Section 120 of CERCLA, DOD and EPA are required to enter into an IAG within 180 days of the completion of EPA's review of the remedial investigation and feasibility study at a site. According to EPA officials, shortly after Section 120 was enacted, EPA and DOD acknowledged that regulatory oversight during the investigation phase was required if EPA was to meet its statutory obligation regarding remedy selection at NPL sites. Beginning in 1988, EPA and DOD agreed to model language for IAGs which included a provision to enter into IAGs earlier than mandated by statute—prior to the remedial investigation stage—to establish the roles and responsibilities of EPA and DOD to investigate and clean up sites. IAGs are required to include, at a minimum, a review of the alternative remedies considered and the selected remedy, a schedule for cleanup, and plans for long-term operations and maintenance. The Federal Facility Compliance Act, among other things, authorized EPA to order the cleanup of contaminated sites by initiating administrative enforcement actions against a federal agency under RCRA, on the same basis as they would be applied to private parties.

Disputes have recently arisen between EPA and DOD regarding the terms of IAGs governing cleanup and whether EPA had a sufficient basis for administrative enforcement actions at several DOD sites. In addition, in recent years, EPA has added fewer sites to the NPL. According to EPA's 2007 annual report on Superfund, more than 75 percent of all sites listed on the NPL—both federal and nonfederal—were listed before 1991. Since fiscal year 2000, EPA added five DOD sites to the NPL (see figure 1).

In this context, we agreed to determine (1) the extent of EPA's oversight during assessment and cleanup at DOD NPL and non-NPL sites and (2) why EPA has proposed fewer DOD sites for the NPL since the early 1990s.

To determine the extent of EPA's oversight during assessment and cleanup at NPL and non-NPL DOD sites, we reviewed EPA policies and documentation on oversight processes, and interviewed officials at EPA headquarters and four regional offices to determine the extent to which the agency helps to ensure that the most contaminated DOD sites are expeditiously assessed and cleaned up. We also reviewed documentation and interviewed DOD officials on the agency's environmental restoration program and efforts to clean up contaminated DOD sites. To determine why EPA has proposed fewer DOD sites for the NPL since the early 1990s, we reviewed EPA's file documentation on contaminated DOD sites and interviewed officials at EPA headquarters and selected EPA regions. We excluded from our review sites under DOD's military munitions response program due to the ongoing uncertainty associated with defining unexploded ordnance as hazardous substances and the fact that GAO has ongoing work in this area.

We conducted work at four EPA regions—Atlanta, Chicago, Dallas, and San Francisco—which, taken together, are responsible for about half of all DOD sites in EPA's database of contaminated federal facilities. We selected the Atlanta and Chicago regions because they are responsible for five DOD sites that EPA proposed for the NPL but which were not listed. We selected the San Francisco region because it has the largest number of contaminated DOD sites. We selected the Dallas region to pretest our review methodology because it was geographically convenient. We conducted this performance audit in accordance with generally accepted government auditing standards between January 2008 and March 2009. Those standards require that we plan and perform the audit to obtain sufficient, appropriate evidence to provide a reasonable basis for our findings and conclusions based on our audit objectives. We believe that the evidence obtained provides a reasonable basis for our findings and conclusions based on our audit objectives. More detail on the scope and methodology of our review is presented in appendix I.

RESULTS IN BRIEF

While EPA evaluates DOD's preliminary assessments of all DOD sites on the Hazardous Waste Compliance Docket, according to EPA officials, the agency has little to no enforceable oversight authority under Section 120 of the cleanup of the majority of these sites because most are not on the NPL. Of the 985 current hazardous release DOD sites, EPA has oversight authority of the 140 DOD sites on the NPL; 11 of these NPL sites do not have IAGs in place that CERCLA Section 120 requires to guide cleanup activity, DOD choosing instead to conduct cleanup with minimal, if any, EPA oversight. The remaining 845 DOD sites are overseen by other cleanup authorities—primarily the states—or required no further action under CERCLA following assessment. Therefore, state agencies or another regulatory authority, rather than EPA, oversee the cleanup of hazardous substance releases at most contaminated DOD sites. Most states have their own cleanup programs to address hazardous waste sites and RCRA corrective action authority to clean up RCRA sites. While EPA regions have some oversight of states' RCRA programs by reviewing site files and providing technical advice to states, EPA defers oversight authority to states for the cleanup of individual RCRA sites. Our review of 389 non-NPL DOD sites at four EPA regions showed that for more than one-half of these sites, EPA generally did not propose to list these sites because it determined that the condition of the sites did not satisfy the criteria to score a high Hazard Ranking System (HRS) score—that is, little to no hazardous release or the potential for a hazardous release was found—or because it deferred the sites to another cleanup program, most often RCRA. EPA regional officials were unable to provide documentation for the agency's decision not to list the remaining sites we reviewed, however, because original site file records were missing or inconclusive. EPA has IAGs with DOD in place for most of its NPL sites—129 of the 140 DOD sites on the NPL. According to an EPA headquarters official, EPA is generally satisfied with the cleanup of DOD NPL sites where there is an IAG. However, the remaining 11 sites do not have IAGs because DOD has disagreed with the terms of the provisions contained in the agreements, stating the terms conflict with or go beyond CERCLA or its regulatory requirements. Despite the CERCLA requirement for IAGs at all NPL federal facility sites, CERCLA Section 120 imposes no specific sanctions if a federal agency refuses to enter into an IAG. Although EPA may initiate administrative enforcement actions, in appropriate circumstances, under other laws, such as RCRA and the Safe Drinking Water Act, to compel DOD to clean up contaminated sites, EPA chose not to pursue enforcement actions until 2007, more than 10 years after these sites were listed on the NPL. In its most recent report to Congress in 2007, EPA noted the number of NPL sites with IAGs but did not explain the basis for the 11 DOD sites without IAGs. Later that year, the agency issued administrative enforcement orders under RCRA and the Safe Drinking Water Act against four of these sites. Each order stated that contamination at the respective sites may present an imminent and substantial endangerment to health or the environment and directed DOD to carry out certain cleanup and related actions. In May 2008, DOD sent a memorandum to the Department of Justice (DOJ) asking DOJ to resolve a dispute over EPA's authority to issue the orders. In December 2008, DOJ issued a letter upholding EPA's authority to issue administrative cleanup orders at DOD NPL sites in appropriate circumstances, and to include in IAGs certain provisions other than those specifically enumerated in CERCLA.[5]

Since the mid-1990s, EPA has listed fewer DOD sites on the NPL than in previous years for three key reasons. First, EPA does not generally list DOD sites that are being addressed under other federal or state programs to avoid duplication of remedial actions. Second, DOD and EPA officials told us that over the years, DOD has discovered fewer hazardous substance releases, making fewer sites available for listing. Fewer sites have been discovered, in part because DOD has been identifying and cleaning up hazardous releases for more than two decades, and because DOD has improved its management of waste generated during its ongoing operations. Finally, in rare instances, EPA did not list some contaminated defense sites due to the objections of other interested parties. For example, although EPA proposed listing five DOD sites between 1994 and 2000, the agency ultimately chose not to complete the listing process for them. At four sites, the states' governors did not support placement of these sites on the NPL. The governors for three of these sites cited the perceived stigma of NPL listing and potential adverse economic effect as the reasons why the state did not support listing. The governor did not support listing the fourth site after it was closed under the Base Realignment and Closure program and DOD began to clean up the site. Although EPA may list sites over the objections of a governor, EPA officials told us they generally do not list federal sites without a governor's concurrence. According to EPA regional officials, EPA did not list the fifth DOD site because DOD objected, and the Office of Management and Budget (OMB) recommended against listing. OMB officials encouraged EPA to defer listing for 6 months to provide DOD with more time to address personnel and contractor changes and demonstrate remediation progress. If after that time, progress was not forthcoming, then listing was to be pursued, but in fact, never was. EPA officials said that cleanup has taken place at all five sites and that it was either unlikely or unclear that the sites would qualify for listing on the NPL based on the current conditions at the sites.

We provided a draft of this chapter to EPA and DOD for review and comment. In general, EPA agreed with the findings and conclusions of our report and supported our suggestion that Congress consider amending CERCLA to expand the agency's enforcement authority. While EPA stated that such authority would help assure timely and protective cleanup, DOD disagreed stating that EPA has sufficient involvement at NPL sites regardless of whether IAGs are in place and should strive to more effectively implement its authority under existing law. Despite DOD's position that EPA is sufficiently involved at DOD NPL sites without IAGs, EPA disagrees. Statutory requirements provide for independent EPA oversight, not a mere

10   United States Government Accountability Office

opportunity for EPA review and comment. Therefore, we assert that expanding EPA's enforcement authority is appropriate to ensure that cleanups are being done properly at federal facility NPL sites.

BACKGROUND

Various environmental statutes, including CERCLA and RCRA, govern the reporting and cleanup of hazardous substances and hazardous waste at DOD sites. Specific provisions in these laws establish requirements for addressing hazardous waste cleanup or management. Key aspects of these requirements for federal facilities are described below:

Comprehensive Environmental Response, Compensation, and Liability Act. The Comprehensive Environmental Response, Compensation, and Liability Act (CERCLA) of 1980 was passed to give the federal government the authority to respond to actual and threatened releases of hazardous substances, pollutants, and contaminants that may endanger public health and the environment. The EPA program under CERCLA is better known as "Superfund" because Congress established a large trust fund that is used to pay for, among other things, remedial actions at nonfederal sites on the NPL.[6] Federal agencies are prohibited from using the Superfund trust fund to finance their cleanups and must, instead, use their own or other appropriations.[7]

Figure 2 depicts the number of NPL sites listed by EPA as of November 2008, which totals 1,587 sites. Of these, 140 were DOD NPL sites, representing the majority of federal facility sites on the NPL. According to EPA's 2007 annual report on Superfund, more than 75 percent of all sites on the NPL—both federal and private—were listed before 1991. Since fiscal year 2000, EPA has added five DOD sites to the NPL.

CERCLA does not establish regulatory standards for the cleanup of specific substances, but requires that long-term cleanups comply with applicable or relevant, and appropriate requirements. These may include a host of federal and state standards that generally regulate exposure to contaminants. The National Oil and Hazardous Substances Pollution Contingency Plan (NCP) outlines procedures and standards for implementing the Superfund program. The NCP designates DOD as the lead agency at defense sites, though as described below, it must carry out its responsibilities consistent with EPA's oversight role under Section 120 of CERCLA, including EPA's final authority

Superfund: Greater EPA Enforcement and Reporting are Needed to... 11

to select a remedial action if it disagrees with DOD regarding the remedy to be selected.[8]

In 1986, the Superfund Amendments and Reauthorization Act (SARA) added provisions to CERCLA specifically governing the cleanup of federal facilities. Under Section 120 of CERCLA, as amended, EPA must take steps that assure completion of a preliminary site assessment by the responsible agency for each site in the Federal Agency Hazardous Waste Compliance Docket.[9] This preliminary assessment is reviewed by EPA, together with additional information, to determine whether the site poses little or no threat to human health and the environment or requires further investigation or assessment for potential proposal to the NPL. SARA also added Section 211 of CERCLA, which established DOD's Defense Environmental Restoration Program providing legal authority governing cleanup activities at DOD installations and properties.

CERCLA Section 120 also establishes specific requirements governing IAGs between EPA and federal agencies. The contents of the IAGs must include at least the following three items: (1) a review of the alternative remedies considered and the selection of the remedy, known as a remedial action; (2) the schedule for completing the remedial action; and (3) arrangements for long-term operations and maintenance at the site. DOD and EPA are required to enter into an IAG within 180 days of the completion of EPA's review of the remedial investigation and feasibility study at a site.

SARA's legislative history explains that, while the law already established that federal agencies are subject to and must comply with CERCLA, the addition of Section 120 provides the public, states, and EPA increased authority and a greater role in assuring the problems of hazardous substance releases at federal facilities are dealt with by expeditious and appropriate response actions.[10] The relevant congressional conference committee report establishes that IAGs provide a mechanism for (1) EPA to independently evaluate the other federal agency's selected cleanup remedy, and (2) states and citizens to enforce federal agency cleanup obligations, memorialized in IAGs, in court.[11] Specifically, the report states that while EPA and the other federal agency share remedy selection responsibilities, EPA has the additional responsibility to make an independent determination that the selected remedial action is consistent with the NCP and is the most appropriate remedial action for the affected facility. The report also observes that IAGs are enforceable documents just as administrative orders under RCRA and, as such, are subject to SARA's citizen suit and penalties provisions. Thus, penalties can be assessed against federal agencies for violating terms of agreements with

EPA.[12] However, at sites without IAGs, EPA has only a limited number of enforcement tools to use in compelling compliance by a recalcitrant agency; similarly, states and citizens also lack a mechanism to enforce CERCLA.

Resource Conservation and Recovery Act. In 1976, Congress passed the Resource Conservation and Recovery Act (RCRA) giving EPA the authority to regulate the generation, transportation, treatment, storage, and disposal of hazardous waste. Under RCRA, EPA may authorize states to carry out many of the functions of the statute in lieu of EPA under a state's hazardous waste programs and laws. Almost all states are authorized to implement some portion of the RCRA program. Forty-eight states are currently authorized to implement the RCRA base program to manage hazardous waste treatment, storage, and disposal. (Only Alaska and Iowa are not authorized to implement the RCRA base program.) Forty-three states are authorized to implement the RCRA corrective action program which expands a state's RCRA authority to include managing the cleanup of releases of hazardous waste and hazardous constituents.

EPA has a policy to defer sites, which are being managed under RCRA, from placement on the NPL, known as the RCRA deferral policy. Where this policy is applied, cleanup proceeds under RCRA, generally through an authorized state corrective action program, rather than CERCLA. EPA regions may defer a federal facility site to RCRA even if the site is eligible for the NPL. In 1996, Congress amended CERCLA to authorize EPA to consider non-CERCLA cleanup authorities when making a listing determination for federal facility sites if the site is already subject to an approved federal or state cleanup plan. According to EPA policy, the criteria to defer a federal facility site from the NPL to RCRA are: (1) the CERCLA site is currently being addressed by RCRA Subtitle C corrective action authorities under an existing enforceable order or permit containing corrective action provisions; (2) the response under RCRA is progressing adequately; and (3) the state and community support deferral of NPL listing. According to EPA, deferral from one program to another is often the most efficient and desirable way to address overlapping requirements, and deferrals to RCRA may free CERCLA oversight resources for use in situations where another authority is unavailable. In these instances, state agencies or another regulatory authority, rather than EPA, oversee the cleanup of hazardous substance releases.

Superfund: Greater EPA Enforcement and Reporting are Needed to... 13

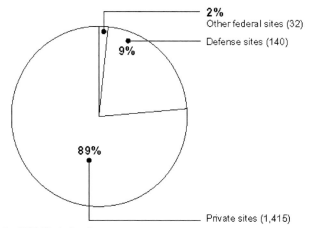

Source: EPA's CERCLA database.
Note: As of November 2008, the total number of federal facilities and private sites on the NPL was 1,587. The 32 other federal NPL sites included 21 Department of Energy sites, 2 Department of Agriculture sites, 1 Federal Aviation Administration site, 1 Coast Guard site, 2 National Aeronautics and Space Administration sites, 1 Small Business Administration site, 2 Department of the Interior sites, 1 Department of Transportation site, and 1 EPA site.

Figure 2. Private, Federal, and DOD Sites on the NPL

Other non-CERCLA cleanup authorities EPA considers in deciding whether to list a site include state cleanup programs (often referred to as voluntary cleanup programs) and DOD's environmental response program. See appendix II for a summary of these cleanup programs.

The NCP provides the methods and criteria for carrying out site discovery, assessment, and cleanup activities under CERCLA. Figure 3 depicts the process by which EPA and federal agencies assess a site for inclusion on the NPL and address contamination at federal NPL sites.

The CERCLA cleanup process is made up of a series of steps, during which specific activities take place or decisions are made. The key steps in this process are included in figure 3.

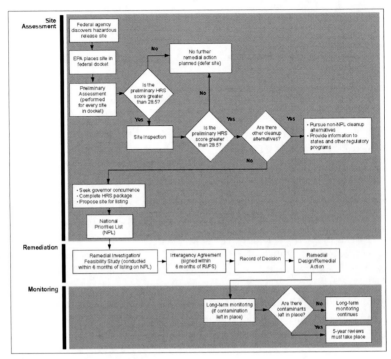

Source: EPA.

Figure 3. Key Stages of the CERCLA Process to Address and Clean Up Hazardous Waste at Federal Facilities

DOCKETREPORTINGCATEGORIES

The categories for listing a facility on the docket provide an initial basis for assessing hazardous waste contamination. The four reporting categories are:

- RCRA Section 3005: facilities for which agencies have applied for an EPA permit for hazardous waste treatment, storage, or disposal.
- RCRA Section 3010: facilities where hazardous materials are generated, transported, treated, stored, or disposed.
- RCRA Section 3016: facilities with hazardous waste activities that federal agencies have reported in their inventories.
- CERCLA Section 103: facilities for which agencies have reported any releases or spills of a hazardous substance.

Every 6 months, EPA is also required to publish in the Federal Register a list of the federal facilities that were added to the docket during the preceding 6-month period and notify regions of any actual or threatened hazardous substance release. At about the same time, EPA also lists these sites in its computerized CERCLA information database, an inventory of actual and potential hazardous releases at federal and private sites. As of October 2008, EPA's CERCLA information database listed 12,621 federal and private sites.

Site discovery. When a federal agency identifies an actual or suspected release or threatened release to the environment on a federal site, it notifies EPA, which then lists the site on its Federal Agency Hazardous Waste Compliance Docket. The docket is a listing of all federal facilities that have reported hazardous waste activities under three provisions of RCRA or one provision of CERCLA. RCRA and CERCLA require federal agencies to submit to EPA information on their facilities that generate, transport, store, or dispose of hazardous waste or that have had some type of hazardous substance release or spill. EPA updates the docket periodically.

Preliminary assessment. The lead agency (DOD, in this case) conducts a preliminary assessment of the site by reviewing existing information, such as facility records, to determine whether hazardous substance contamination is present and poses a potential threat to public health or the environment. EPA regions review preliminary assessments to determine whether the information is sufficient to assess the likelihood of a hazardous substance release, a contamination pathway, and potential receptors. EPA regions are encouraged to complete their review of preliminary assessments of federal facility sites listed in EPA's CERCLA database within 18 months of the date the site was listed on the federal docket. EPA may determine the site does not pose a significant threat to human health or the environment and no further action is required. If the preliminary assessment indicates that a long-term response may be needed, EPA may request that DOD perform a site inspection to gather more detailed information.

Site inspection. The lead agency (DOD, in this case) samples soil, groundwater, surface water, and sediment, as appropriate, and analyzes the results to prepare a report that describes the contaminants at the site, past waste handling practices, migration pathways for contaminants, and receptors at or near the site. EPA reviews the site inspection report and, if it determines the release poses no significant threat, EPA may eliminate it from further consideration. If EPA determines that hazardous substances, pollutants, or contaminants have been released at the site, EPA will use the information collected during the preliminary assessment and site inspection to calculate a preliminary HRS score.

HRS scoring. If EPA determines that a significant hazardous substance release has occurred, the EPA region prepares an HRS scoring package. EPA's HRS assesses the potential of a release to threaten human health or the environment by assigning a value to factors related to the release such as (1) the likelihood that a hazardous release has occurred; (2) the characteristics of the waste, such as toxicity and the amount; and (3) people or sensitive environments affected by the release.

National Priorities List. If the release scores an HRS score of 28.50 or higher, EPA determines whether to propose the site for placement on the NPL. CERCLA requires EPA to update the NPL at least once a year.

Governor's concurrence. Before placing a site on the NPL, the EPA Regional Administrator sends a written inquiry to the governor seeking a written response from the state addressing whether it will support a listing decision. According to EPA regional officials, EPA usually contacts the governor before calculating the HRS score due to the high cost and length of time required to prepare a scoring package. If EPA calculates an HRS score of 28.50 or higher and the governor agrees with EPA to list the site, the site is eligible for inclusion on the NPL. However, where the governor does not support listing, but the EPA region firmly believes listing is necessary, a process, involving OMB for federal facilities, is followed before a listing decision is made.

Following the decision to place a site on the NPL, several steps lead to the selection of a cleanup remedy and its long-term operation and maintenance. These steps are described below:

Remedial investigation and feasibility study. Within 6 months after EPA places a site on the NPL, the lead agency (DOD, in this case) is required to begin a remedial investigation and feasibility study to assess the nature and extent of the contamination. The remedial investigation and feasibility study process includes the collection of data on site conditions, waste characteristics, and risks to human health and the environment; the development of remedial

Superfund: Greater EPA Enforcement and Reporting are Needed to... 17

alternatives; and testing and analysis of alternative cleanup methods to evaluate their potential effectiveness and relative cost. EPA, and frequently the state, provide oversight during the remedial investigation and feasibility study and the development of a proposed plan, which outlines a preferred cleanup alternative. After a public comment period on the proposed plan, EPA and the federal facility sign a record of decision that documents the selected remedial action cleanup objectives, the technologies to be used during cleanup, and the analysis supporting the remedy selection.

Interagency agreement. Within 6 months of EPA's review of DOD's remedial investigation and feasibility study, CERCLA, as amended, requires that DOD enter into an IAG with EPA for the expeditious completion of all remedial action at the facility. (EPA's policy however, is for federal facilities to enter into an IAG after EPA places the site on the NPL.) The IAG is an enforceable document that must contain, at a minimum, three provisions: (1) a review of remedial alternatives and the selection of the remedy by DOD and EPA, or remedy selection by EPA if agreement is not reached; (2) schedules for completion of each remedy; and (3) arrangements for the long-term operation and maintenance of the facility.

> **DERP**
>
> Under the Defense Environmental Restoration Program, DOD must ensure that EPA and appropriate state and local officials have notice and an adequate opportunity to comment on (1) the discovery of releases or threatened releases of hazardous substances at a facility; (2) the extent of the threat to public health and the environment which may be associated with any such release or threatened release; (3) DOD proposals to carry out response actions with respect to any such release or threatened release; and (4) the initiation of any response action with respect to such release or threatened release and the commencement of each distinct phase of such activities. The Defense Environmental Restoration Program does not require DOD to respond to the comments or provide a mechanism for EPA or other officials to compel a response or other action if they believe DOD's proposed activities are inadequate.

Remedial design and remedial action. During the remedial design and remedial action process, the lead agency (DOD, in this case) develops and implements a permanent remedy on the site as outlined in the record of decision and IAG.

Monitoring. Long-term monitoring occurs at every site following construction of the remedial action. This includes the collection and analysis of data related to chemical, physical, and biological characteristics at the site to determine whether the selected remedy meets CERCLA objectives to protect human health and the environment. For NPL or non-NPL sites where hazardous substances, pollutants, or contaminants were left in place above levels that do not allow for unlimited use and unrestricted exposure, every 5 years following the initiation of the remedy, the lead agency (DOD, in this case) must review its sites. The purpose of a 5-year review, similar to long-term monitoring, is to assure that the remedy continues to meet the requirements contained in the record of decision and is protective of human health and the environment.

Federal Facility Compliance Act. The Federal Facility Compliance Act of 1992, which amended RCRA, authorizes EPA to order the cleanup of contaminated sites by initiating administrative enforcement actions against a federal agency under RCRA, including the imposition of fines and penalties. The act authorizes EPA to initiate administrative enforcement actions against federal agencies in the same manner and under the same circumstances as actions would be initiated against a person.

Enforcement. Several factors hinder the enforcement of cleanup requirements at federal facilities. DOJ has taken the position that EPA may not sue another federal agency to enforce cleanup requirements. EPA may not issue cleanup orders under CERCLA to other federal agencies without DOJ's concurrence. EPA may issue cleanup orders to other federal agencies under RCRA and the Safe Drinking Water Act, but not all RCRA orders can provide for administrative penalties. IAGs also generally contain administrative penalty provisions. Third parties, such as states and citizens groups, may sue to enforce IAGs and administrative orders under the "citizen suit" and other public participation provisions of CERCLA, RCRA, and Safe Drinking Water Act, but such litigation can be time consuming.[13]

EPA EVALUATES ALL POTENTIALLY CONTAMINATED DOD SITES FOR LISTING, BUT DOES NOT OVERSEE CLEANUP AT MOST HAZARDOUS WASTE DOD SITES

While EPA oversees and evaluates DOD's preliminary assessments of all DOD sites suspected of having a hazardous release, the agency has little to no oversight of the cleanup of most of these sites because most are not on the NPL. EPA reviews DOD sites to determine whether to propose placement on the NPL. However, only 140 of the 985 current DOD sites with hazardous waste appear on the NPL. EPA and DOD have not finalized IAGs for the remaining 11 sites, which impedes EPA's ability to enforce cleanup, such as approving detailed cleanup schedules and applying administrative penalties. EPA only recently began using enforcement action at DOD NPL sites where an IAG is not in place. State agencies, rather than EPA, oversee the cleanup of hazardous waste at most DOD sites.

EPA Reviews DOD Sites to Determine whether to Propose NPL Listing

DOD performs preliminary assessments of all federal DOD sites on the Federal Agency Hazardous Waste Compliance Docket. EPA regions review the assessments to determine whether releases pose a threat to human health and the environment and if so, whether hazardous substances are being released into the environment. DOD's preliminary assessments are based on readily available and historical data of suspected releases on DOD sites. DOD reports the results of preliminary assessments to EPA, which often requests additional information such as data on site geography, prior activities at the site, and the source and destination of the hazardous release. According to EPA guidance, EPA regions should complete their review of preliminary assessments within 18 months of when the site was listed on the federal docket; however, EPA officials from two regions told us that DOD may take 2 to 3 years to complete a preliminary assessment because EPA does not have an independent authority under CERCLA to enforce a time line for completion of the preliminary assessment. Based on their review of the preliminary assessment, EPA regional officials may determine that no further action is needed at the site or request that DOD perform a more comprehensive site

20   United States Government Accountability Office

inspection by sampling groundwater and other media on site. Following DOD's investigation, EPA regional officials may: determine that no further action is needed at the site; defer the site to another regulatory authority, such as a state agency, for cleanup; or begin the process to propose the site for placement on the NPL.

Few Hazardous Waste DOD Sites Considered for Listing Are Ultimately Placed on the NPL

Of the 985 DOD sites contaminated with hazardous substances, EPA placed 140 sites—about 15 percent—on the NPL; the remaining 845 sites are generally overseen by a cleanup authority other than EPA. Sites on the NPL are considered among the most dangerous of all hazardous substance sites, based on the evaluation criteria used by EPA. EPA may propose to list sites that (1) have an HRS score of 28.50 or higher; (2) a state designates as its top priority, regardless of the HRS score; or (3) are subject to a health advisory issued by the Agency for Toxic Substances and Disease Registry and meet certain other criteria.[14] In practice, however, few sites meet these criteria. Further, even if a site is eligible for placement on the NPL based on the HRS score, EPA may choose to defer the site to RCRA. As we discuss later in this chapter, our review of non-NPL DOD sites in four EPA regions demonstrated that available data supporting these decisions is limited. EPA regional officials were unable to provide a rationale for EPA's decision to not list almost one-half of the 389 sites that we reviewed because site file documentation was inconclusive or missing. For the remaining sites, EPA did not propose listing because officials determined the sites did not satisfy the criteria to score a high HRS score or deferred them to another regulatory authority.

More than a Decade after Listing, 11 DOD NPL Sites Do Not Have IAGs, Impeding EPA's Ability to Enforce Cleanup Actions at Those Sites

Although EPA has IAGs in place with DOD for 129 of the 140 DOD sites on the NPL, IAGs have not been finalized at the remaining 11 sites remaining. According to an EPA headquarters official, EPA is generally satisfied with the cleanup of DOD NPL sites where DOD has signed IAGs. EPA has encountered few problems at these sites, the EPA official said, because DOD is held accountable for compliance with the provisions of the IAGs and if differences arise, the agreements provide EPA with an enforceable process to address the issue. EPA and DOD have not finalized IAGs for the remaining 11 DOD NPL sites, however. As a result, DOD has been cleaning up 11 sites without IAGs, inhibiting EPA's ability to seek enforcement actions that compel attention to schedules and milestones. Under CERCLA, as amended, EPA and DOD must enter into negotiated IAGs for the expeditious completion of all necessary remedial action at each DOD site on the NPL. IAGs must include, at a minimum, the alternative remedies and the selected remedy, a schedule for completing the remedial action, and arrangements for long-term operation and maintenance of the facility. According to EPA, the schedule is enforceable and often found in a site management plan that documents and provides for re-evaluation of schedules and priorities for cleanup. In addition, EPA officials indicated that IAGs generally also include consultative provisions that document time frames for review and comment on documents by each agency as well as administrative penalties for DOD's failure to comply with the agreed-upon cleanup tasks and milestones. The IAG therefore documents EPA's expectations of DOD, and provides for administrative penalties against the department when it does not comply with the activities agreed to in the document. Without the IAG, EPA does not have the needed criteria, or a foundation upon which an enforcement action may be taken, and has limited ability to sanction DOD without going to court, which DOJ does not allow it to do. The 11 DOD NPL sites—2 Army, 2 Navy, and 7 Air Force facilities—were placed on the NPL at least a decade ago, between 1994 and 1999, except for 1 of the Air Force sites, which was listed in 1983.[15] As of early March 2009, however, DOD has not finalized IAGs for any of these sites. [16] In its most recent report to Congress for fiscal year 2007, EPA indicated the number of NPL sites with IAGs and facilities where EPA had issued enforcement orders. However, EPA's report did not clearly indicate that there were 11 DOD NPL sites without IAGs and the reasons why.

There is a long history of EPA and DOD efforts to negotiate IAGs, beginning in 1988. Key actions taken by these agencies are listed in table 1.

Although CERCLA requires that federal agencies enter into IAGs with EPA to govern the cleanup of NPL sites within 180 days of EPA's review of the remedial investigation and feasibility study, DOD officials told us they have not finalized IAGs for 11 NPL sites because DOD disagreed with some of the terms of the provisions contained in the agreements. DOD also indicated they feel that EPA has adequate authority through its remedy selection process and that the IAG serves primarily as an administrative roadmap.[17] Although the Defense Environmental Restoration Program statute requires DOD to take actions that provide EPA with adequate opportunity to review and comment at key phases of cleanup, there are no formal ramifications when DOD does not comply. Without an IAG, EPA lacks a documentation roadmap that demonstrates review and comment on key decisions. An IAG would identify areas of concern at a site and the process being used to address them. At DOD NPL sites without IAGs, such as at Langley Air Force Base in Maryland, DOD did not obtain EPA concurrence before signing a unilateral record of decision that identifies the remedial action. As a result, according to EPA, the agency cannot confirm whether all areas of contamination have been identified or whether they are being addressed properly. In 1988 and supplemented in 1999 and 2003, DOD and EPA developed model language for specific provisions representing the most contentious issues encountered in earlier negotiations. Although DOD agreed to the model language, it has disagreed with some of the specific terms contained in the provisions of agreements based on these models, such as those that, in DOD's opinion, conflict with or go beyond CERCLA or its regulatory requirements. DOD officials also stated that EPA has been unwilling to negotiate the terms of these provisions with DOD.

Although EPA has some oversight of the cleanup of NPL sites where DOD has not entered into an IAG, EPA officials told us the agency has only limited ability to carry out cleanup enforcement actions at federal facilities. For example, at sites where DOD has entered into an IAG, EPA has the authority to approve and modify a sites' sampling plan. In contrast, at NPL sites without an IAG, although DOD may send copies of draft plans and reports to EPA, it is often without regard to schedule or a process for vetting issues back and forth as defined in IAG provisions. Therefore, EPA's role is limited to reviewing many plans after they are finalized without the opportunity to provide input to the cleanup process. According to EPA headquarters officials, EPA is not seeking excessive enforcement authority at DOD NPL sites but intends to hold DOD to the same enforceable oversight it has at private sites. In fact, federal agencies are more often subject to much less stringent enforcement provisions. DOJ has taken the position that EPA may not sue another federal agency to enforce cleanup requirements, which effectively restricts EPA's ability to compel compliance through civil judicial litigation. According to EPA, enforcement provisions contained in the agreements, such as stipulated penalties, are generally less onerous for federal facilities than they are for private parties. The terms of the provisions,

Superfund: Greater EPA Enforcement and Reporting are Needed to... 23

regardless of whether they are based on model language agreed upon between DOD and EPA, are necessary for EPA to carry out its role to enforce the cleanup process, EPA officials said. The IAG is not simply an administrative document but an essential tool, without which EPA and the states cannot assure the public that DOD is properly identifying and addressing hazardous waste at contaminated DOD sites.[18]

Table 1. Chronology of Events to Negotiate IAGs for DOD NPL Sites

Date	Event
1987–1988	Following the passage of SARA in 1986, DOD finalized IAGs with EPA at 4 NPL sites.
June 1988	To facilitate negotiation of additional IAGs, EPA and DOD approved a model agreement that included: • Standard language for 11 provisions—such as dispute resolution, enforcement, and stipulated penalties—to address fines for failure to submit certain documents or comply with the terms and conditions of the agreement. • A list of 27 other provisions—such as remedial action, site access, and transfer of property—where the specific terms were left to be negotiated for each site.
1989–1998	One hundred-three DOD NPL sites finalized IAGs with EPA.
February 1999	EPA and DOD agreed to modify the model agreement in light of changes to DOD's budget and increasing costs of operations to include: • Modified provisions for deadlines (near-term milestones) and funding. • New provisions for a site management plan, budget development, and scheduling.
1999–2003	Twelve DOD NPL sites finalized IAGs with EPA.
October 2003	EPA and DOD agreed to the following: Modified the model agreement to add provisions for institutional and engineering controls to ensure that contaminants do not pose an unacceptable risk to human health or the environment at sites where contamination is left in place. • Established a "dual-track" approach whereby EPA and DOD allow the military services to negotiate land use control provisions for sites with EPA following one of two approaches[a]
2004–2008	Ten DOD NPL sites finalized IAGs with EPA.

24 United States Government Accountability Office

Table 1. (Continued)

Date	Event
July–November 2007	EPA issued administrative cleanup orders to four DOD NPL sites that did not have IAGs.
December 2007	While the military services were allowed to continue to negotiate IAGs with EPA, the Office of the Secretary of Defense directed that they must follow the model agreement, and any additional provisions added to the IAG must first be approved by OSD and the other services. Further, any changes to the provisions of the model IAG would be allowed only through negotiations between OSD and EPA.
May 2008	DOD asked DOJ and OMB to resolve a dispute between DOD and EPA over the terms of the IAGs and the circumstances under which EPA may issue administrative orders.
December 2008	DOJ issued a letter upholding EPA's authority to issue administrative cleanup orders at DOD NPL sites in appropriate circumstances, and to include in IAGs certain provisions other than those specifically enumerated in CERCLA.
January 2009	Eleven DOD NPL sites do not have IAGs.
February 2009	The Deputy Under Secretary of Defense notifies EPA that DOD is willing to accept the latest IAG for Fort Eustis in Virginia as the new model for the remaining DOD NPL sites without IAGs and instructs the military services to begin negotiations with EPA.
March 2009	On March 4, the Navy signed IAGs for the Naval Air Station Whiting Field in Florida and the Naval Computer Telecommunication Area Administrative Master Station in Hawaii. Since EPA also signed these IAGs, the next steps required before the agreements are effective include acquiring the states' signatures and completing a public comment period and EPA review.

Source: DOD and GAO's analysis of relevant documents and interviews with agency personnel.

[a]The dual-track approach is a set of two principles for negotiating land use control provisions. Based on Navy and Air Force principles, the Navy's approach was to negotiate terms beyond the model agreement while the Air Force's approach was to add language to a record of decision without changing the language of the provision in the agreement.

EPA Only Recently Used Enforcement Action at DOD NPL Sites without IAGs

Although EPA may initiate enforcement actions to compel the cleanup of contaminated sites, EPA only recently began to use this authority at DOD NPL sites without IAGs. In 2007, EPA issued four administrative cleanup orders— three under RCRA and one under the Safe Drinking Water Act[19]—to four DOD NPL sites—Tyndall Air Force Base in Florida, McGuire Air Force Base in New Jersey, Air Force Plant 44 in Arizona, and Fort Meade in Maryland— that do not have IAGs. The orders stated that an imminent and substantial endangerment from contamination may be present on the sites and required DOD to notify EPA of its intent to comply with the orders and clean up. The Air Force did not agree with EPA's assertion that an imminent and substantial endangerment existed at Air Force Plant 44, but agreed to perform the work required by the order. At the remaining two Air Force sites and one Army site, the services disagreed with EPA's assertion that an imminent and substantial endangerment existed and indicated that the failure to enter into an IAG at the site was an inappropriate basis for issuing an order. The Air Force also argued that compliance with the orders would not accelerate study and cleanup but, rather, that the additional paperwork required for compliance would delay implementation of ongoing investigation and cleanup. The Air Force and Army did not notify EPA of their intent to comply with the orders within the time frame required and stated they would continue to clean up these sites under their CERCLA removal and lead agency authority. According to DOD, some of these sites are nearly cleaned up. For example, as of July 2008, DOD estimated that three of the four sites had cleaned up about two-thirds or more of the contamination on site. According to EPA headquarters officials, DOD's estimation of the cleanup at these sites is inconsistent with EPA's assessment and there is still much work to be performed at each of these sites. For example, according to EPA headquarters officials, Tyndall Air Force Base has not completed a single record of decision for work to be performed and McGuire Air Force Base has not completed a single investigation.

In May 2008, DOD requested that DOJ and OMB resolve the disagreement between DOD and EPA as to the basis upon which EPA may issue imminent and substantial endangerment orders under RCRA and the Safe Drinking Water Act, and the terms of federal facility agreements regarding cleanup at DOD NPL sites. As of November 2008, OMB was noncommittal regarding its involvement. On December 1, 2008, DOJ issued a letter upholding EPA's authority to issue administrative cleanup orders at DOD NPL

26　United States Government Accountability Office

sites in appropriate circumstances. Specifically, the letter stated, among other things, that

- EPA may issue imminent and substantial endangerment orders to DOD in accordance with RCRA and the Safe Drinking Water Act;
- EPA may issue such orders at a site even if it would not have done so had there been an IAG under CERCLA for the site; and
- while IAGs are consensual undertakings, and DOD is not necessarily required to agree to all IAG terms EPA seeks beyond those enumerated in CERCLA, EPA may require DOD to agree in an IAG to follow EPA guidelines, rules, and criteria in the same manner, and to the same extent as these apply to private parties.[20]

As of early March 2009, the Air Force and Army did not have IAGs for these four sites, including the site being cleaned up under the Safe Drinking Water Act order. [21]

State Agencies Oversee the Cleanup of Hazardous Waste at Most DOD Sites

Because the majority of contaminated DOD sites are not on the NPL, most DOD site cleanups are overseen by state agencies rather than EPA, as allowed by CERCLA. CERCLA provides that state cleanup and enforcement laws apply to federal facilities not included on the NPL. Under CERCLA, EPA may choose to defer a federal facility site to another cleanup authority, such as RCRA, even though the site is eligible for placement on the NPL. Of the 845 DOD sites not on the NPL, EPA generally determined that no further action was needed at the sites either because (1) the sites did not have hazards that would score high enough for NPL listing or (2) EPA deferred oversight of DOD's response at the sites to the states or other regulatory authorities. Most states have their own cleanup programs to address hazardous waste sites and RCRA corrective action authority to clean up RCRA sites. While EPA regions have some oversight of states' RCRA programs by reviewing site files and providing technical advice to the state, EPA defers oversight authority to states for the cleanup of non-NPL RCRA sites. EPA does not exercise day-to-day oversight of state cleanup programs but has entered into memorandums of understanding or agreement with some states. For example, EPA and the state of Ohio entered into a memorandum of agreement that defined the roles and responsibilities of EPA and the state for non-RCRA cleanups.

EPA PROPOSES FEW CONTAMINATED DOD SITES BASED ON EPA POLICY AND DOD'sMATURINGINVENTORY OF HAZARDOUSWASTE SITES

Since the 1990s, EPA has proposed fewer DOD sites for the NPL than in previous years for three key reasons. First, EPA defers the majority of DOD sites to other statutory authorities for cleanup under state oversight, and to avoid duplicating efforts, it does not list these sites. Second, over the years, DOD has discovered fewer hazardous substance releases, resulting in fewer sites for assessment and potential proposal for the NPL. Third, state officials or other federal agencies may, on occasion, object to EPA's proposal to list contaminated DOD sites, and while EPA can still propose listing the site, it usually does not. Based on our review of 389 unlisted DOD sites from four EPA regions, we found EPA did not list about half of these sites because EPA determined that little to no hazardous release had occurred or it deferred the site to a state for oversight, often because a contamination response was already underway.

EPA Does Not List DOD Sites That Are Cleaned Up under RCRA or Other Programs

In 1996, Congress amended CERCLA to specify that a response under another cleanup authority is an appropriate factor to consider when making a determination whether to list a federal site.[22] Since then, EPA has generally not proposed listing contaminated DOD sites that are being cleaned up under other federal or state programs. Under EPA's deferral policy, it may choose to defer sites to RCRA, even if sites are eligible for the NPL, where (1) the CERCLA site is currently being addressed by RCRA Subtitle C corrective action authorities under an existing enforceable order or permit containing corrective action provisions, (2) the response is progressing adequately, and (3) the state supports deferral of placement on the NPL. According to EPA headquarters officials, during the early years of CERCLA, the Superfund program was the primary means by which EPA assured that contamination at federal facilities was assessed and cleaned up. In recent years, however, other cleanup programs such as RCRA have evolved and matured so that placement on the NPL is just one of several tools available to address contamination. EPA policy allows regions to defer a federal facility site to RCRA even though the site is eligible for the NPL. Officials from two EPA regions said that almost all of the region's DOD sites were being cleaned up under RCRA at the time they were assessed and to avoid adding unnecessary and redundant regulatory oversight, the regions chose to leave them under RCRA for cleanup. EPA regions also defer sites from the NPL that are being cleaned up under a state cleanup program. EPA headquarters officials said that many sites proposed for placement on the NPL were referred to EPA by the states but that, over the years, states developed their own cleanup programs and did not refer as many sites to EPA. As a result, EPA headquarters officials said that EPA is not proposing to list as many sites based on states' referrals.

DOD Is Identifying Fewer Contaminated Sites

DOD is discovering and reporting fewer new or additional hazardous substance releases because, over the years, many potentially contaminated waste sites have been identified and cleaned up and waste management practices have changed. Discovery of new DOD sites has been infrequent, making fewer sites available to EPA for assessment and proposal for inclusion on the NPL. According to Army officials, beginning in the early 1980s, the Army conducted initial assessments to identify potentially contaminated sites. As a result, Army officials said, the Army's installation restoration program inventory is mature and, for the most part, complete. According to a Navy official, during the 1980s and 1990s, the Navy also conducted assessments to identify and catalog the majority of contaminated Navy sites. DOD officials also stated that because of controls placed on the management of hazardous materials and wastes as a result of well-established laws, there are relatively fewer releases or threats of release, and operational releases are immediately addressed. EPA officials generally agreed that DOD has identified fewer contaminated DOD sites in recent years because, EPA officials said, the services have a fairly well-inventoried universe of sites, and old or abandoned DOD sites are no longer being discovered. Further, EPA headquarters officials said, DOD has cleaned up hazardous waste sites over the years, has tremendous cleanup efforts underway, and has the budgets to fund them.

States May Object to EPA's Proposal to List Contaminated DOD Sites

EPA policy recommends states' governors to be included in the decision whether to list sites on the NPL and, in cases where a state does not agree that EPA should list a site, EPA's policy recommends that a region work closely with the state to resolve the state's concerns. If the region is unable to resolve the state's concerns and EPA believes it has sufficient reasons to proceed with listing, EPA may list the site on the NPL without the state's concurrence; however, according to EPA headquarters officials, EPA will not list a site without agreement from the state.[23]

On rare occasions, EPA proposed but ultimately did not list some contaminated DOD sites. Four sites were not listed because the states' governors did not support listing. EPA did not list a fifth site because OMB recommended against listing. Although these five sites were not listed, EPA regional officials said that all five sites are being cleaned up, have a remedy in place that is protective of human health and the environment, or the site has been cleaned up to the point that it no longer meets the requirements for placement on the NPL. Specifically:

Rickenbacker Air National Guard Base. In 1994, DOD closed the remaining portions of the Rickenbacker Air National Guard Base in Lockbourne, Ohio, which had been in use since 1942 providing aircraft refueling operations. Fuel contamination and chemical releases were found around underground fuel lines and tanks and near former storage areas and buildings. Trichloroethylene (TCE) has been found in soil and near groundwater.[24] In January 1994, EPA proposed placing the site on the NPL but did not do so because the governor did not agree, citing the stigma that NPL listing would have on current, planned, and future economic development as well as the potential to adversely affect the economic development of adjacent sites. The governor also proposed that the Ohio EPA oversee investigation and cleanup activities at the site under the state's cleanup program. Today, portions of the site are being cleaned up under RCRA while other portions are being cleaned up under CERCLA and DOD's Base Realignment and Closure program, with state oversight. According to EPA headquarters officials, EPA and the Air Force agreed the site should be cleaned up for commercial-industrial use. The Air Force transferred portions of the facility to another state agency for cleanup and signed an agreement with the state to clean up the remaining lands, in accordance with CERCLA. However, the Air Force has

30   United States Government Accountability Office

refused to include land use restrictions in its selected remedy, as EPA would normally do for sites on the NPL. Nonetheless, cleanup at the site is proceeding, EPA regional officials said, and the site no longer meets the requirements for the NPL.

Air Force Plant 85. Air Force Plant 85 in Columbus, Ohio, manufactured and tested aircraft and missile systems between 1941 and 1994. Wastes produced from these operations included acids from metal cleaning and electroplating, cyanide wastes, and paint strippers. From 1984 to 1990, the Air Force identified multiple sources of potential hazardous waste contamination, including two nearby streams and a creek. TCE and other chlorinated solvents were found in groundwater; polychlorinated biphenyls (PCB),[25] solvents, and metals were found in soil; and various metals and solvents were found in sediment. In January 1994, EPA proposed placing the site on the NPL but did not do so because the governor objected, again citing the stigma of listing and its potential effects on economic development. The governor also proposed that the Ohio EPA oversee investigation and cleanup activities at the site under the state's cleanup program. Air Force Plant 85 is being cleaned up under Ohio's Voluntary Cleanup Program which, according to EPA officials, follows the CERCLA process. According to EPA regional and Air Force officials, the Air Force has cleaned up or has a remedy in place at 11 of the 13 sources of hazardous substances releases at the site and is expected to have all remedies in place by 2011.

Arnold Engineering Development Center. The Arnold Engineering Development Center near Tullahoma and Manchester, Tennessee, is an Air Force test and research organization that simulates flight conditions in ground-test facilities. The site contains contaminated landfills, leaching pits, and testing areas. Jet and rocket fuels, solvents, and other shop wastes have been detected in the main testing area. PCBs also have been detected in soil samples collected in the main testing area and in wastewater and surface water runoff in a retention reservoir. In August 1994, EPA proposed placing the site on the NPL but did not do so because the governor did not concur. EPA regional officials said that state officials told them Tennessee preferred to clean up the site under a state cleanup program and speculated that many states may prefer this arrangement because of the perception of a stigma associated the NPL. Further, the Arnold Engineering Development Center was competing with a DOD facility in another state to install a wind tunnel and the Tennessee governor's office was concerned that NPL listing would hurt the site's

Superfund: Greater EPA Enforcement and Reporting are Needed to... 31

chances. The Air Force is cleaning up the Arnold Engineering Development Center under RCRA with EPA and state oversight. EPA regional officials said that Air Force actions to date on the site are protective of human health and control the migration of contaminated groundwater. While Air Force officials said they expect all remedies to be in place by the end of fiscal year 2011, EPA regional officials indicated the goal for final construction of the remedy is 2020.[26]

Wurtsmith Air Force Base. Wurtsmith Air Force Base, a 5,000-acre site near Oscoda, Michigan, has performed various air support missions since it was established in the early 1920s, such as aircraft and vehicle maintenance and air refueling. In 1977, the Air Force sampled drinking water and monitoring wells on the site and found solvents, including TCE. The U.S. Geological Survey also sampled and found TCE in the groundwater. The base closed in June 1993 and in January 1994, EPA proposed placing the site on the NPL. However, EPA did not list the site because the state did not support listing after DOD placed the site in the Base Realignment and Closure program and progressed with cleanup under state oversight. Although TCE is still present in groundwater plumes, EPA regional officials said the site has been cleaned up to the point that it would no longer meet the requirements for the NPL.

Chanute Air Force Base. Chanute Air Force Base in Rantoul, Illinois, provided military and technical training for Air Force and civilian personnel on the operation and maintenance of military aircraft and ground support equipment until DOD closed the base in 1990. The primary sources of hazardous waste on the site include various landfills, fire training areas and buildings that contained oil-water separators, underground storage tanks, and sludge pits. The primary concern was the potential for this contamination to migrate into a nearby creek. In April 2000, the governor wrote to the EPA region to express his support for placing Chanute Air Force Base on the NPL, citing the state's concern about past operation and disposal practices at the site and because the state was unable to reach an agreement with the Air Force on how the site should be cleaned up. In December 2000, EPA proposed placing the site on the NPL but the Air Force objected, citing a perception that listing was a stigma and argued it could clean up the site by 2005 and on schedule if it did not have to suspend cleanup to negotiate the provisions of an IAG. The Air Force asked OMB to mediate the dispute. EPA presented its case for listing the site to OMB, pointing out that the site's HRS score supported a proposal for listing, the governor of the state concurred, and listing would help to assure that DOD would enter into an IAG with EPA to clean up. In 2003, OMB determined that EPA should not proceed with listing. OMB encouraged EPA to defer listing the site for 6 months to provide DOD with time to address personnel and contractor changes and demonstrate remediation progress. If, after that time, progress was not forthcoming, then listing was to be pursued, but in fact, never was. Although EPA officials told us that cleanup at Chanute has progressed slowly, milestones were met and EPA did not list the site. The Air Force estimates that it will have all remedies in place by the end of fiscal year 2012 and all property transferred from Air Force control by the end of fiscal year 2014. Although cleanup is behind schedule, according to EPA regional officials, the site has been cleaned up to the point that it is unclear whether the site would score for the NPL if the listing process was started today. For example, three of the four landfills have been capped and are no longer active. Remedial investigation reports of the creek do not show the levels of contamination detected when EPA proposed listing the site. Despite the slow progress to clean up, EPA regional officials said they believe that proposing the site for listing ultimately helped to start the cleanup process.

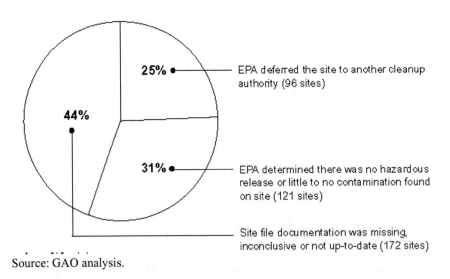

Source: GAO analysis.

Figure 4. Findings from GAO's Review of 389 DOD Sites

Superfund: Greater EPA Enforcement and Reporting are Needed to... 33

Four EPA Regions Did not List Sites Due to a Lack of Contamination or Hazardous Release or Because Sites Were Deferred to Another Cleanup Authority

As part of our review, we asked officials from four EPA regions to provide the primary basis for their decision to not propose placing 389 DOD sites under their jurisdiction on the NPL. (See figure 4.) Based on a review of site records and interviews with EPA regional officials, we found EPA did not propose listing almost one-third of these sites (121 of 389, or 31 percent) because site assessments found little to no contamination or hazardous release on the site or no contamination exposure pathway or receptor. In instances where EPA scored these sites, the HRS score was below the minimum hazard ranking threshold for the NPL. One-quarter of these sites (96 of 389, or 25 percent) were not proposed for the NPL because EPA deferred them to another authority, such as a state agency under its RCRA authority. We were unable to determine the rationale for EPA's decision to not list less than half of these sites (172 of 389, or 44 percent) because site file records were missing, inconclusive, or not up to date. For example, some site files showed that EPA had not yet determined whether to propose listing, even though the site assessment was conducted decades ago. According to EPA region officials, record-keeping practices have varied over the years so that, in some cases, site files and the basis for EPA's decisions were not well documented or maintained.

CONCLUSIONS

While the number of DOD sites considered for placement on the NPL has declined over the past decade, DOD sites still account for 9 percent of all NPL sites. Despite years of negotiations, DOD and EPA have not finalized IAGs to clean up 11 of the 140 DOD NPL sites. Most are more than a decade overdue, yet EPA has made few efforts to use its enforcement authority under CERCLA to compel parties to enter into IAGs, and to select remedies at sites without agreements. While the Federal Facility Compliance Act authorizes EPA to apply the same RCRA enforcement policies to federal facilities as it does to nonfederal facilities, EPA has not taken enforcement action at most federal sites. In light of prolonged disagreements between DOD and EPA over the terms of the IAGs, and the absence of any statutory consequences for failing to enter into an IAG, now may be the time to reconsider the provisions required by CERCLA for effective EPA oversight. While the law offers accountability through citizen suits, transparency through public participation provisions, legal recourse through enforceable schedules, and mechanisms for addressing conflicts through dispute resolution provisions, at sites without IAGs EPA lacks the leverage needed to provide strong environmental stewardship. Bringing the parties together for further discussions with relevant oversight committees may facilitate resolution at the sites without IAGs. While the pattern of delays in DOD's preliminary assessment process appeared to go unchallenged by EPA, we believe EPA's failure to enforce a time line for completion further exacerbated this process. These conditions suggest a need for stronger enforcement and reporting as well as a serious commitment to address ongoing challenges.

We believe Congress should be kept apprised of the situations where agreements are lacking. However, EPA has not used its annual report to Congress to provide this information.[27] Moreover, because EPA was unable to make available documentation of the basis for its decisions whether to list or not list DOD sites, it is impossible for EPA to provide a justification for its decisions for many of the sites placed on or left off of the NPL.

MATTER FOR CONGRESSIONAL CONSIDERATION

Given the critical nature of Superfund cleanup for protecting public health, and the long-term commitment necessary to maintain strong environmental stewardship at federal facilities, we encourage Congress to ensure accountability by DOD and EPA by raising concerns about the impasse between these federal agencies, if IAGS are not finalized within 60 days following issuance of this chapter. Specifically, Congress should consider amending CERCLA Section 120 to authorize EPA to impose administrative penalties at federal facilities placed on the NPL that lack IAGs within the CERCLA-imposed deadline of 6 months after completion of the remedial investigation and feasibility study. This leverage could help EPA better satisfy its statutory responsibilities with agencies that are unwilling to enter into agreements where required under CERCLA Section 120. In addition, Congress may wish to consider amending Section 120 to authorize EPA to require agencies to complete preliminary assessments within specified time frames.

Recommendations for Executive Action

To facilitate congressional oversight of the Superfund program and provide greater transparency to the public on the cleanup of DOD sites, we recommend that the Administrator of EPA improve its record keeping in the following manner. Consistent with good management practices defined in EPA's Superfund program implementation manual and to ensure that meaningful data are available for the agency's reports to Congress, EPA should establish a record-keeping system, consistent across all regions, to accurately document EPA decisions regarding the proposal of DOD sites for inclusion or exclusion on the NPL and the basis for each decision.

APPENDIX I. OBJECTIVES, SCOPE, AND METHODOLOGY

We were asked to determine (1) the extent of the Environmental Protection Agency's (EPA) oversight during assessment and cleanup at Department of Defense (DOD) National Priority List (NPL) and non-NPL sites and (2) why EPA has proposed fewer DOD sites for inclusion on the NPL since the early 1990s.

To examine the extent of EPA's oversight during assessment and cleanup of DOD NPL and non-NPL sites, we reviewed the Comprehensive Environmental Response, Compensation, and Liability Act (CERCLA) and other legislation governing the cleanup of federal hazardous waste sites, as well as EPA Superfund program policy and guidance, to determine the roles and responsibilities of EPA and federal agencies, such as DOD, to implement the CERCLA process and assess and clean up hazardous waste. We reviewed EPA and DOD reports to the Congress on the Superfund and Defense Environmental programs, respectively. We reviewed EPA and DOD policy and guidance on interagency agreements (IAG), including the model agreements, and correspondence relating to the negotiation of IAGs for selected DOD sites. We conducted several interviews with EPA and DOD headquarters officials on issues related to IAGs and enforcement. At GAO's request, EPA provided data from its computerized CERCLA information database of actual and potential hazardous releases at federal and private sites. Based on these data, we worked with EPA to identify the universe of DOD sites and obtain certain information on these sites, such as NPL status. To determine the reliability of the CERCLA information database, an EPA

36   United States Government Accountability Office

headquarters official contacted each EPA region and asked them to verify selected information, such as the number of DOD sites and their NPL status. During site visits to selected EPA regions, we also confirmed certain information in the CERCLA information database by reviewing site file documentation, where available, and interviewing EPA region officials. Based on this work, we determined that these data were sufficiently reliable for the purposes of this chapter. We interviewed EPA headquarters officials on the agency's policies and processes under the Superfund program to ensure that contaminated federal DOD sites, both NPL and non-NPL, are assessed and cleaned up. We interviewed DOD headquarters officials on DOD's role and responsibilities to identify, report, assess, and clean up, as necessary, hazardous releases at NPL and non-NPL DOD sites. We also interviewed officials at four EPA regions on their oversight of contaminated federal DOD sites, both NPL and non-NPL, to assure that sites are assessed and cleaned up. We conducted our work at four EPA regions—Atlanta, Chicago, Dallas, and San Francisco—which, taken together, were responsible for about half of the 845 non-NPL DOD sites. We selected the Atlanta and Chicago regions because they were responsible for five DOD sites that EPA proposed for NPL inclusion but which were not listed. We selected the San Francisco region because it had the largest number of non-NPL DOD sites. We selected the Dallas region to pretest our review methodology because it was geographically convenient.

To determine why EPA has proposed fewer DOD sites for NPL inclusion since the early 1990s, we reviewed EPA policy and guidance on proposing sites for the NPL and interviewed EPA headquarters and regional officials on the reasons why EPA has proposed fewer sites. We interviewed DOD headquarters officials on its progress to identify and assess potentially contaminated DOD sites and the reasons why fewer hazardous releases have been identified. We interviewed EPA and DOD officials on contaminated DOD sites that EPA proposed for the NPL, why some were not listed, and the status of cleanup at these sites. Finally, for selected DOD sites, we evaluated the basis for EPA's decision to not propose listing certain contaminated DOD sites by reviewing site file documentation and interviewing EPA regional officials regarding all non-NPL DOD sites at four EPA regions. We excluded from our review sites under DOD's military munitions response program because of the ongoing uncertainty concerning the degree to which spent military munitions are subject to RCRA and CERCLA, and the fact that GAO has ongoing work in this area. Based on our review of contaminated DOD sites at four EPA regions, we attempted to determine the primary basis for

EPA's decision to not propose to list the site. However, we were unable to confirm the basis for EPA's decision to not propose listing less than one-half of the sites surveyed (172 of 389, or 44 percent) because site file documentation, such as records of EPA's decisions and recommendations concerning sites, was missing or inconclusive. For example, officials at one EPA region told us they could not determine how many sites required no further action after either a preliminary assessment or site inspection because, prior to 1990, the region did not document the basis for determining that no further action was required.

APPENDIX II.OTHERCLEANUP PROGRAMS

In addition to the Comprehensive Environmental Response, Compensation, and Liability Act (CERCLA), there are a number of other cleanup authorities EPA considers in deciding whether to list a site include state cleanup programs (often referred to as voluntary cleanup programs) and the Defense Environmental Restoration program. Specifically:

State Cleanup Programs. Over the years, most states have developed their own cleanup programs, often referred to as voluntary state cleanup programs. Some state cleanup programs address hazardous waste sites independent of a state's Resource Conservation and Recovery Act (RCRA) program. Often, state cleanup projects begin with a preliminary site assessment and if contamination is suspected, an on-site investigation is conducted. EPA does not have oversight of state cleanup programs but has entered into memoranda of agreement or understanding with some states, recognizing the use of the state's cleanup program to address hazardous waste sites under a state's non-RCRA authority.

Defense Environmental Restoration Program. In 1986, Congress amended CERCLA and required that DOD establish an environmental restoration program under which all response actions at hazardous waste contaminated sites—such as site identification, investigation, and cleanup—must be conducted consistent with Section 120 of CERCLA.[28] More than 15 years later, the National Defense Authorization Act for Fiscal Year 2002 required that DOD also develop an inventory of all DOD sites known or suspected to contain unexploded ordnance, military munitions, or munitions constituents throughout the United States and develop a methodology for prioritizing response actions at these sites. Today, DOD's environmental response program includes an installation restoration program, which in 1985 began addressing hazardous releases resulting from past practices, and a military munitions response program, established as a separate program in 2001, to address safety and environmental hazards from unexploded ordnance and munitions on other-than-operational ranges (ranges that are closed, transferred or transferring). As of fiscal year 2007, DOD reported there were 27,950 installation restoration program sites on DOD facilities and former defense sites, of which 23,980, or 86 percent, had achieved "remedy in place" or "response complete" status[29]. At 3,537 munitions response sites at current DOD facilities and former defense sites, a total of 940, or 27 percent, had achieved "remedy in place" or "response complete" status. DOD completed an initial inventory of munitions response sites in fiscal year 2002. Since then, DOD has been working to reconcile its inventory which includes conducting site assessments (preliminary assessments and, if needed, site inspections) of all sites. DOD estimates it will complete site assessments for all munitions response sites by the end of fiscal year 2010 except for sites on former defense sites. Former defense sites represent the majority of sites with suspected munitions response sites and, according to DOD, site assessments for munitions sites on former defense sites will not be completed until about 2013.

Under the Defense Environmental Restoration Program, DOD cleans up environmental hazards and contamination on active installations, installations being closed under DOD's Base Realignment and Closure program, and at formerly used defense sites. DOD is required to carry out response cleanup actions under the program, subject to, and in a manner consistent with, Section 120 of CERCLA. DOD is required to report annually to Congress on its environmental restoration programs. As of fiscal year 2007, DOD reported that its goal was to clean up all known releases (or achieve a "remedy in place" status) on active installations by the end of fiscal year 2014 and all sites on formerly used defense sites by the end of fiscal year 2020.

APPENDIX III. COMMENTS FROM THE ENVIRONMENTAL PROTECTION AGENCY

UNITED STATES ENVIRONMENTAL PROTECTION AGENCY
WASHINGTON, D.C. 20460

FEB 2 4 2009

Mr. John B. Stephenson
Director
Natural Resources and Environment
U.S. Government Accountability Office
Washington, DC 20548

Re: EPA comments on the Government Accountability Office's (GAO) draft report to Congress entitled *Greater EPA Oversight, Enforcement, and Reporting Are Needed to Enhance Defense Site Cleanup (GAO-09-278)*

Dear Mr. Stephenson:

Thank you for the opportunity to review GAO's draft report entitled *Greater EPA Oversight, Enforcement, and Reporting Are Needed to Enhance Defense Site Cleanup (GAO-09-278)*. We appreciate GAO's review of EPA's role in the clean up and restoration of contaminated Department of Defense (DoD) properties.

EPA agrees that effective use of strong enforcement will help to assure timely and protective clean up of federal facilities. As for GAO's draft findings related to the National Priorities List (NPL), we offer in this letter general comments on two of the key reasons cited in the draft report for the slowing pace of DoD site listings: (1) States may object to EPA's proposals to list sites; and (2) DoD has discovered fewer hazardous substance releases (DoD's "mature inventory" of sites). In addition to comments on these two issues, you will find detailed comments in the enclosure to this letter.

Concerning the pace of NPL listings, GAO's draft report indicates that State objections to NPL listing may be one of four causes for the slowing pace of NPL listings of DoD sites. This conclusion was based on an examination of four instances in which States opposed the placement of DoD sites on the NPL during the period 1994-2001. It should be noted, however, that 44 federal sites *were* placed on the NPL (without State objection) during the same time period, including 37 DoD sites. Subsequent to that time frame, only three DoD sites were proposed for the NPL, and none of these were opposed by the States. In fact, of these three sites, one was listed at the specific request of a Governor, and the other two were listed with State support. We don't believe, therefore, that one can conclude that formal State objections have significantly contributed to the reduction in NPL listings.

GAO also cited "DoD's mature inventory" of contaminated properties and the discovery of fewer sites to explain the reduction in NPL listings. While typical sources of contamination on DoD bases have been fairly well characterized, as noted in the draft report, other areas on these bases have not been adequately characterized, and there are still some newly discovered sources of contamination (e.g., munitions response sites under DoD's Military Munitions Response Program). While some slowing in the pace of DoD site listings is to be expected, the total number of potential NPL sites in DoD's inventory, and hence, the degree to which the pace of listing is explained by the "mature inventory" of sites, cannot be determined while there remains insufficient or missing site specific information to support decision making.

Again, thank you for the opportunity to comment. Please contact me if I can be of assistance, or your staff may call Bobbie Trent in EPA's Office of the Chief Financial Officer at 202.566.0983.

Sincerely,

Barry N. Breen
Acting Assistant Administrator, OSWER

APPENDIX IV. COMMENTS FROM THE DEPARTMENT OF DEFENSE

OFFICE OF THE UNDER SECRETARY OF DEFENSE
3000 DEFENSE PENTAGON
WASHINGTON, DC 20301-3000

ACQUISITION
TECHNOLOGY
AND LOGISTICS

FEB 2 0 2009

Mr. John B. Stephenson
Director, Natural Resources and Environment
U.S. Government Accountability Office
441 G Street, N.W.
Washington, DC 20548

Dear Mr. Stephenson:

This is the Department of Defense (DoD) response to the GAO draft report, "SUPERFUND: Greater EPA Oversight, Enforcement, and Reporting Are Needed to Enhance Defense Site Cleanup," dated January 27, 2009 (GAO Code 360916/GAO-09-278).

The Department acknowledges receipt of the draft report. Even though there are no recommendations directed at DoD, we have concerns with some of GAO's factual statements, findings, conclusions and recommendation. Detailed technical comments on the report were provided separately.

In summary, DoD has major concerns with the following issues:

- The title of the report is misleading and not based on the facts presented in the report. The facts presented in the report do not support the conclusion that the DoD Cleanup program is not meeting the requirements of the Comprehensive Environmental Response, Compensation, and Liability Act (CERCLA) or the Resource Conservation and Recovery Act (RCRA). The report does not provide evidence that additional Environmental Protection Agency (EPA) oversight or enforcement authorities is needed and if provided these authorities will speed up cleanup or lead to better decision making. According to the findings of the report EPA does not need more or new oversight enforcement authorities, EPA needs to more effectively implement their authorities they already have under existing law.

- Congress has defined the term "defense site" in 10 U.S.C. Section 2710(e)(1). The report definition of defense site conflicts with this legal definition and implies Department of Energy sites are defense sites. Recommend using the term DoD site or DoD facility.

Superfund: Greater EPA Enforcement and Reporting are Needed to... 41

- The number of DoD National Priorities List (NPL) sites and remaining Federal Facilities Agreements (FFA) to be signed is inaccurate. DoD has 140 not 141 NPL site and 11 not 12 FFA remaining to be signed. The Middlesex Sampling Plant is a site under the Formerly Utilized Sites Remedial Action program (FUSRAP). It is assigned by law to the U.S. Army Corps of Engineers Civil Works Program, as the lead agency for the conduct of remedial actions at certain sites associated with the former Manhattan Engineer District and Atomic Energy Commission. It is not a under the direction or control of DoD cleanup program. The Middlesex Sampling Plant property is owned by the United States and under the accountability of the Department of Energy.

The facts presented in the report do not support the conclusions and recommendations. EPA is actively involved in review of response actions at DoD NPL sites, regardless of whether an FFA has been signed. The report indicates no evidence that the lack of an FFA at any of the sites has delayed, diminished, or reduced the timeliness or quality of the response actions taken by DoD with EPA, state and public involvement, at any DoD NPL site. DoD has made significant progress which is demonstrated in the report for the five sites EPA decided to not list on the NPL.

Sincerely,

Wayne Arny
Deputy Under Secretary of Defense
(Installations and Environment)

CC:
OGC (E&I)
DASA (ESOH)
DASN (E)
DASAF (EESOH)

End Notes

[1] Executive Order 12580 directs the responsible federal agency to carry out the preliminary assessment. This executive order also delegates certain CERCLA authorities to the Department of Defense.

[2] The Hazard Ranking System (HRS) is the principal mechanism EPA uses to place sites on the NPL. The HRS serves as a screening device to evaluate the potential for releases of uncontrolled hazardous substances to cause human health or environmental damage. The HRS provides a measure of relative rather than absolute risk. It is designed so that it can be consistently applied to a wide variety of sites. [40 C.F.R. Pt. 300, App. A, § 1.0]

[3] The NPL is composed of 157 final and 15 deleted federal sites and 1,100 final and 315 deleted private sites.

42 United States Government Accountability Office

[4]For the purposes of this review, both NPL and non-NPL DOD sites are federal facilities where DOD is the agency responsible for the cleanup of hazardous waste resulting from past practices.

[5]The letter stated that "because an interagency 'agreement' denotes a consensual undertaking, we do not think that DOD necessarily is required to agree to all extra-statutory terms demanded by EPA. We think that EPA nonetheless may require DOD to agree in the IAG to follow, 'in the same manner and to the same extent' as they apply to private parties, any 'guidelines, rules, regulations, and criteria' established by EPA and made applicable to non-federal facilities under CERCLA." The letter also noted that whether the facts identified in each order present a sufficient basis to support EPA's finding of an imminent and substantial endangerment is a factual issue that DOJ was unable to address.

[6]This trust fund was financed primarily by taxes on crude oil and certain chemicals, as well as an environmental tax assessed on corporations based upon their taxable income. Although the authority for these taxes expired in 1995, the trust fund continued to receive revenue from various other sources, including appropriations from the general fund. EPA receives annual appropriations from the trust fund for program activities; since 1981, Superfund appropriations have totaled over $32 billion in nominal dollars, or about $1.2 billion annually.

[7]GAO, *Superfund: Funding and Reported Costs of Enforcement and Administration Activities,* GAO-08-841R (Washington, D.C.: July 18, 2008).

[8]Under the NCP, DOD maintains its lead agency responsibilities whether the remedy is selected by DOD for non-NPL sites or by EPA and the federal agency or by EPA alone for NPL sites under CERCLA Section 120. Executive Order 12580, Superfund Implementation (Jan. 23, 1987) as amended delegates certain presidential authorities under CERCLA to the Secretary of Defense. Specifically, the executive order provides that CERCLA response authorities "are delegated to the Secretaries of Defense and Energy, with respect to releases or threatened releases where either the release is on or the sole source of the release is from any facility or vessel under the jurisdiction, custody or control of their departments, respectively, including vessels bare-boat chartered and operated. These functions must be exercised consistent with the requirements of Section 120 of the Act."

[9]Executive Order 12580 delegates to DOD the authority for carrying out preliminary assessments and site inspections at DOD sites. CERCLA imposes no deadlines for completing preliminary assessments.

[10]H.R. Rep. No. 99-253, pt. 1 at 95 (1985).

[11]H.R. Conf. Rep. No. 99-962 at 242 (1986).

[12]In technical comments on our report, DOD asserted that sufficient EPA oversight can occur without an IAG, so long as a signed record of decision exists for a given site. This view is inconsistent with the language in CERCLA Section 120 and the legislative history. First, Section 120 uses the term "interagency agreement," not the term "record of decision" which appears elsewhere in SARA; a reference in Section 120 to IAGs *instead of* records of decision is far more than a semantic accident. Second, as indicated above, the IAG serves to provide a basis for enhanced EPA cleanup oversight as well as enforcement by states and citizens. DOD failed to explain how a record of decision would serve a similar purpose, and in particular failed to address the IAG's role in enhancing state and citizen enforcement activities. While the conference committee report states that a record of decision signed by *both* EPA and the other federal agency can serve as an IAG, H.R. Conf. Rep. No. 99-962 at 242, we read this to mean that the terms of the record of decision may also be used as the terms of the IAG if *both* parties agree and are otherwise consistent with CERCLA. To the extent the conference report can be read to suggest that an IAG is not required at a DOD NPL site with a co-signed record of decision, this reading is inconsistent with the language of the statute, which provides for no such exception.

[13]DOD recently asserted that a state's decision to sue to enforce compliance with a cleanup order could result in the state losing certain DOD grant funds. Recently, an organization of state

Superfund: Greater EPA Enforcement and Reporting are Needed to... 43

waste management officials criticized DOD's position as being inconsistent with statutes, such as RCRA, that authorize states to bring such enforcement actions.

[14] The Agency for Toxic Substances and Disease Registry, part of the Department of Health and Human Services, performs specific functions concerning the effect on public health of hazardous substances in the environment such as public health assessments of waste sites, response to emergency releases of hazardous substances, and education and training concerning hazardous substances.

[15] The 11 DOD NPL sites without IAGs include (1) Air Force Plant 44, Arizona; (2) Andrews Air Force Base, Maryland; (3) Brandywine Defense Reutilization and Marketing Office Salvage Yard, Maryland; (4) Fort Meade, Maryland; (5) Hanscom Field, Massachusetts; (6) Langley Air Force Base, Virginia; (7) McGuire Air Force Base, New Jersey; (8) Naval Air Station Whiting Field, Florida; (9) Naval Computer Telecommunication Area Administrative Master Station, Hawaii; (10) Redstone Arsenal, Alabama; and (11) Tyndall Air Force Base, Florida. A twelfth NPL site, Middlesex Sampling Plant, New Jersey, also does not have an IAG. Middlesex is listed in EPA's CERLCA information database as a Department of Energy site even though the Fiscal Year 1998 Energy and Water Appropriations Bill transferred management of the site to the Army Corps of Engineers. While EPA officials said that the agency considers Middlesex to be a DOD NPL site for the purposes of enforcement and negotiation of IAGs, we excluded it from our list of DOD sites without IAGs.

[16] On Mar. 4, 2009, the Navy began the process for finalizing IAGs at two of its sites. The Navy signed IAGs for the Naval Air Station Whiting Field in Florida and the Naval Computer Telecommunication Area Administrative Master Station in Hawaii. Since EPA has also signed the IAGs, the next steps will be to obtain the states' signatures followed by a public comment period and EPA final review. At the conclusion of this process, the IAGs will be considered effective.

[17] As discussed in the background section of this chapter, SARA's legislative history suggests that IAGs serve primarily as a tool for EPA oversight and as the primary cleanup enforcement mechanism at DOD NPL sites.

[18] This view is consistent with the portions of SARA's legislative history that discuss the IAG provision.

[19] The Safe Drinking Water Act and its amendments established standards and treatment requirements for the nation's drinking water supply and delegated primary implementation and enforcement authority to the states.

[20] The letter is available at http://www.fedcenter.gov/_kd/go.cfm?destination=ShowItem&Item_ID=11085. DOD has also asked OMB to review the terms of the IAGs regarding cleanup at these sites. An executive order provision implementing CERCLA Section 120 directs OMB to facilitate resolution of disputes between EPA and DOD; Executive Order 12580, § 10(a). As of November 2008, OMB has been noncommittal regarding its role with DOD and EPA.

[21] On Dec. 12, 2008—almost 1 year after the effective date of the administrative order—the Army submitted to EPA its notice to comply with the order at Fort Meade, Maryland. On Dec. 23, 2008, the State of Maryland filed suit against the Army seeking to compel the Army's compliance with EPA's administrative order at Fort Meade.

[22] Pub. L. No. 104-201, Div. A, Title III, §§ 330, 110 Stat. 2484 (1996).

[23] According to EPA officials, in the history of the Superfund program, EPA has not listed any site without a state's concurrence. In 1998, EPA proposed listing Fox River in Wisconsin, a private site, without the governor's consent. However, EPA did not finalize the listing because the state and EPA reached an agreement after which cleanup began in 2000.

[24] TCE is a nonflammable, colorless liquid used mainly as a solvent to remove grease from metal but which also is found in adhesives, paint removers, typewriter correction fluids, and spot removers. TCE can cause nervous system effects, liver and lung damage, abnormal heartbeat, coma, and possibly death.

[25]PCBs are a family of chemicals that were used in hundreds of industrial and commercial applications such as electric and hydraulic equipment; as plasticizers in paints, plastics, and rubber products; and in pigments and dyes. PCBs were banned in 1979 and have been demonstrated to cause cancer and effect human immune, reproductive, and nervous systems.

[26]DOD and EPA use different terminology to track cleanup status. DOD tracks the status of cleanup in terms of a "remedy in place" (where the selected remedy is in place and operating) followed by "response complete" (where the required remedial action or operations have been completed.) EPA tracks final construction, or "construction complete," which considers when all physical construction at a site is complete, all immediate threats have been addressed, and all long-term threats are under control. While long-term cleanup actions may still be operating, the site is often ready for another use.

[27]Although a CERCLA requirement for reporting IAG status information was repealed in 2002, DOD reports on the status of NPL sites without IAGs in its annual report to Congress on the Defense Environmental Restoration Program. DOD's report provides a list of the DOD sites without IAGs; it does not provide information on the reasons why IAGs have not been finalized.

[28]10 U.S.C. § 2700 *et seq.* The program is known as the Defense Environmental Restoration Program.

[29] DOD and EPA use different terminology to track cleanup status. DOD tracks the status of cleanup in terms of a "remedy in place" (where the selected remedy is in place and operating) followed by "response complete" (where the required remedial action or operations have been completed.) "Response complete" may also indicate a site was administratively closed; that is, the site did not meet the eligibility criteria for funding under the program, no information was found suggesting that contamination was present, or the property was transferred or is being cleaned up as part of another site. EPA tracks final construction, or "construction complete," which considers when all physical construction at a site is complete, all immediate threats have been addressed, and all long-term threats are under control. While long-term cleanup actions may still be operating, the site is often ready for another use.

In: Contaminated Department of Defense Site...   ISBN: 978-1-61122-465-8
Editors: Douglas B. Ferro   © 2011 Nova Science Publishers, Inc.

Chapter 2

SUPERFUND: INTERAGENCY AGREEMENTS AND IMPROVED PROJECT MANAGEMENT NEEDED TO ACHIEVE CLEANUP PROGRESS AT KEY DEFENSE INSTALLATIONS

United States Government Accountability Office[*]

WHY GAO DID THIS STUDY

Before the passage of federal environmental legislation in the 1970s and 1980s, Department of Defense (DOD) activities contaminated millions of acres of soil and water on and near DOD sites. The Environmental Protection Agency (EPA) has certain oversight authorities for cleaning up contaminants on federal property, and has placed 1,620 of the most contaminated sites— including 141 DOD installations—on its National Priorities List (NPL). As of February 2009, after 10 or more years on the NPL, 11 DOD installations had not signed the required interagency agreements (IAG) to guide cleanup with EPA. GAO was asked to examine (1) the status of DOD cleanup of hazardous substances at selected installations that lacked IAGs, and (2) obstacles, if any, to cleanup at these installations. GAO selected and visited three installations,

---

[*] This is an edited, reformatted and augmented edition of a United States Government Accountability Office publication, Report GAO-10-348, dated July 2010.

reviewed relevant statutes and agency documents, and interviewed agency officials.

WHAT GAO RECOMMENDS GAO

GAO is recommending, among other things, that EPA and DOD identify options that would provide a uniform method for reporting cleanup progress at the installations and allow for transparency to Congress and the public. EPA and DOD agreed with the recommendations directed at them. GAO is also suggesting that Congress may want to consider giving EPA certain tools to enforce CERCLA at federal facilities without IAGs. DOD disagreed with this suggestion. GAO believes EPA needs additional authority to ensure timely and proper cleanup at such sites.

WHAT GAO FOUND

EPA and DOD use different terms and metrics to report cleanup progress; therefore, the status of cleanup at Fort Meade Army Base, McGuire Air Force Base (AFB), and Tyndall AFB is unclear. EPA reports that cleanup at all three installations is in the early investigative phases, while DOD's data suggest that cleanup is further along and, in some cases, in mature stages. EPA and DOD have differing interpretations of cleanup progress because they describe and assess cleanup differently. In particular, while both agencies divide installations into smaller cleanup projects, DOD divides them into units generally smaller than EPA's; therefore, DOD measures its progress in smaller increments. Further, because DOD did not obtain EPA's approval for key cleanup decisions, EPA does not recognize them. Unless key cleanup decisions are justified, documented, and available to the public for review and comment, they are not sufficient under the Comprehensive Environmental Response, Compensation and Liability Act (CERCLA), and once an IAG is in place, some DOD cleanup work may have to be redone. When an agency refuses to enter into an IAG and cleanup progress lags, because of statutory and other limitations, EPA cannot take steps—such as issuing and enforcing orders—to compel CERCLA cleanup as it would for a private party.

A variety of obstacles have delayed cleanup progress at these installations. First, DOD's persistent failure to enter IAGs, despite reaching agreement with

EPA on the basic terms, has made managing site cleanup and addressing routine matters challenging at these installations. For example, in the absence of IAGs, DOD may fund work at other sites ahead of these NPL sites. Second, DOD failed to disclose some contamination to EPA and the public in a timely fashion, including lead shot on a playground, delaying cleanup and putting human health at risk. Third, the extensive use of performance-based contracts at these installations has created pressure to operate within price caps and fixed deadlines. In some cases, these pressures may have contributed to installations not exploring the full range of cleanup remedies, or relying on nonconstruction remedies, such as allowing contaminated groundwater to attenuate over time rather than being cleaned up. In particular, Tyndall AFB's long-standing lack of full compliance with environmental cleanup requirements, such as notification of hazardous releases and EPA's 2007 administrative order, has been an obstacle to verifiable cleanup of that installation.

ABBREVIATIONS

AFB	Air Force Base
ATSDR	Agency for Toxic Substances Disease Registry
CERCLA	Comprehensive Environmental Response, Compensation, and Liability Act
CERCLIS	Comprehensive Environmental Response, Compensation, and Liability Information System
DERP	Defense Environmental Restoration Program
DOD	Department of Defense
DOJ	Department of Justice
E.O.	executive order
EPA	Environmental Protection Agency
IAG	interagency agreement
NCP	National Oil and Hazardous Substances Pollution Contingency Plan
NPL	National Priorities List
OU	operable unit
PA/SI	preliminary assessment and site inspection
PBC	performance-based contract

PCB   polychlorinated biphenyls
PCE   tetrachloroethylene
RCRA   Resource Conservation and Recovery Act
RI/FS   remedial investigation and feasibility study

Source: EPA.

Lead Shot on School Playground at Tyndall Air Force Base in June 2009

ROD   record of decision
SARA   Superfund Amendments and Reauthorization Act
SDWA   Safe Drinking Water Act
SMP   site management plan
TCE   trichloroethylene
VOC   volatile organic compound

July 15, 2010

The Honorable Frank R. Lautenberg
Chairman
Subcommittee on Transportation Safety,
Infrastructure Security, and Water Quality
Committee on Environment and Public Works
United States Senate

The Honorable Robert Menendez
United States Senate

The Honorable Bill Nelson
United States Senate

The Honorable Benjamin L. Cardin
United States Senate

Before federal environmental legislation was enacted in the 1970s and 1980s regulating the generation, storage, treatment, and disposal of hazardous waste, Department of Defense (DOD) activities and industrial facilities contaminated millions of acres of soil and water on and near DOD properties in the United States and its territories. DOD properties released hazardous substances to the environment primarily through industrial operations to repair and maintain military equipment, and the manufacturing and testing of weapons at ammunition plants and proving grounds. From 1986 to 2008, DOD spent $29.8 billion on environmental cleanup and restoration activities at its properties in response to such hazardous releases.[1] Furthermore, in its most recent annual report to Congress, DOD expressed its commitment to full and sustained compliance with federal, state, and local environmental laws and regulations that protect human health and preserve natural resources.

To address the cleanup of hazardous releases at both private and government facilities nationwide, in 1980, Congress passed the Comprehensive Environmental Response, Compensation, and Liability Act (CERCLA), better known as "Superfund."[2] Under CERCLA, as amended, the Environmental Protection Agency (EPA) has certain oversight authorities for cleaning up releases of hazardous substances, pollutants, or contaminants on federal properties. As of April 2010, 1,620 Superfund sites were on EPA's National Priorities List (NPL), which identifies some of the most seriously contaminated sites in the nation, of which 141 or almost 9 percent were DOD properties.[3] As of February 2009, 11 of these properties did not have an interagency agreement (IAG)[4] despite CERCLA's requirement that federal agencies enter into IAGs with EPA within a certain time frame to clean up sites on the NPL.[5] DOD and EPA signed IAGs for 7 of these installations between March 2009 and January 2010, but as of June 2010, DOD had not signed IAGs for 4 of these properties, even though they are required under CERCLA.

You asked us to review activities at selected DOD installations on the NPL that lacked IAGs as of February 2009. Accordingly, this chapter examines (1) the status of DOD cleanup of hazardous substances at selected

50   United States Government Accountability Office

DOD installations that lacked IAGs and (2) obstacles, if any, to progress in cleanup at these selected installations and the causes of such obstacles.

To select installations for review,[6] from the 11 without IAGs at the start of our review, we focused on the 4 that received EPA administrative cleanup orders—Air Force Plant 44 in Arizona, Fort Meade Army Base in Maryland,[7] McGuire Air Force Base (AFB) in New Jersey, and Tyndall AFB in Florida.[8] EPA and DOD agreed that one of these—Air Force Plant 44[9]—was near completion of the ordered cleanup and we therefore eliminated it from our selection of installations. To determine the status of DOD cleanup of hazardous substances at the three remaining installations, we reviewed numerous technical documents and interviewed officials from DOD, EPA, the Agency for Toxic Substances and Disease Registry (ATSDR)—created by CERCLA to help determine the public health consequences of the worst hazardous waste sites—and DOD contractors. To identify any obstacles to progress in cleanup at the selected installations and the causes of such obstacles, we reviewed federal contracting guidelines and technical documents developed by DOD installations and EPA regions, and interviewed officials from DOD, EPA, ATSDR, the Fish and Wildlife Service, and the Architect of the Capitol as well as state officials from Florida, Maryland, and New Jersey. We also reviewed relevant laws, regulations, and policies. Appendix I includes additional information about our selection criteria, scope, and methodology. We conducted this performance audit from January 2009 to July 2010 in accordance with generally accepted government auditing standards. Those standards require that we plan and perform the audit to obtain sufficient, appropriate evidence to provide a reasonable basis for our findings and conclusions based on our audit objectives. We believe that the evidence obtained provides a reasonable basis for our findings and conclusions based on our audit objectives.

BACKGROUND

This section discusses key aspects of relevant laws and history related to the implementation of Superfund and the reporting and cleanup of hazardous substances and hazardous waste at DOD installations.

Relevant Laws and Executive Orders

Resource Conservation and Recovery Act

In 1976, Congress passed the Resource Conservation and Recovery Act (RCRA), establishing requirements, as well as giving EPA regulatory authority, for the generation, transportation, treatment, storage, and disposal of hazardous waste.[10] Section 7003 authorizes EPA to issue administrative cleanup orders where an imminent and substantial endangerment to health and the environment may exist;[11] if a nonfederal recipient fails to comply, EPA can enforce the order, including fines, by requesting that the Department of Justice (DOJ) file suit in federal court. RCRA also authorizes citizen and state suits, including those to enforce an administrative cleanup order.

Comprehensive Environmental Response, Compensation, and Liability Act

The passage of CERCLA in 1980 gave the federal government the authority to respond to actual and threatened releases of hazardous substances, pollutants, and contaminants that may endanger public health and the environment. EPA's program implementing CERCLA is better known as "Superfund" because Congress established a trust fund that is used to pay for, among other things, remedial actions at nonfederal installations on the NPL. Federal agencies cannot use the Superfund trust fund to finance their cleanups and must, instead, use their own or other appropriations.

CERCLA does not establish regulatory standards for the cleanup of specific substances, but requires that remedial actions—which are long-term cleanups—comply with "applicable or relevant and appropriate requirements."[12] These requirements may include a host of federal and state standards that generally regulate exposure to contaminants. CERCLA also establishes authorities for removals, including expeditious response actions by EPA and DOD to reduce dangers to human health, welfare, or the environment such as an emergency response required within hours or days to address acute situations involving actual or potential threat to human health, the environment, or real or personal property due to the release or threatened release of a hazardous substance.[13] Generally, removals are quicker, short-term responses to reduce risks, while remedial actions are the culmination of the full CERCLA process to provide long-term protection of human health and the environment.

The National Oil and Hazardous Substances Pollution Contingency Plan (NCP) outlines procedures and standards for implementing the Superfund program. The NCP designates DOD as the lead agency for cleanup at defense installations. CERCLA requires DOD to comply with the law and the NCP to the same extent as a nonfederal entity; thus, the same process and standards for cleanup apply. Where there has been a release of a hazardous substance where DOD is the lead agency, CERCLA section 103 requires DOD to report such releases above reportable quantities as soon as it has knowledge of such release to the National Response Center,[14] and section 111(g) requires DOD to notify potentially injured parties of such release, and promulgate regulations pertaining to notification. In addition, DOD must carry out its responsibilities consistent with EPA's oversight role under section 120 of CERCLA, including EPA's final authority to select a remedial action at NPL installations if it disagrees with DOD's proposed remedy.

CERCLA section 120 establishes specific requirements governing IAGs between EPA and federal agencies. The contents of the IAGs must include at least the following three items: (1) a review of the alternative remedies considered and the selection of a remedial action by the agency head and EPA (or, if unable to reach agreement, selection by EPA); (2) the schedule for completing the remedial action; and (3) arrangements for long-term operations and maintenance at the installation.[15] Federal agencies and EPA are required to enter into an IAG within 180 days of the completion of EPA's review of the remedial investigation and feasibility study (RI/FS) at an installation. An RI/FS is performed at the site, typically after a site is listed on the NPL. The RI serves as the mechanism for collecting data to characterize site conditions; determine the nature of the waste; assess risk to human health and the environment; and conduct treatability testing to evaluate the potential performance and cost of the treatment technologies that are being considered. The FS is the mechanism for the development, screening, and detailed evaluation of alternative remedial actions. Because such study culminates in a record of decision (ROD), EPA has interpreted this requirement as triggered by the first ROD at an NPL site with multiple cleanup activities. EPA and federal agencies often enter IAGs earlier so the agreement may guide the study process as well.

IAGs between EPA and DOD[16] include a site management plan, which is an annually amended document providing schedules and prioritization for cleanup of the installation, addressing all response activities and associated documentation, as well as milestones. IAGs also specify requirements for documents throughout the cleanup process, addressing DOD's submission, EPA's review, and DOD's response to EPA's comments. For "primary" documents, such as the site management plan, RI/FS work plans and reports, RODs, final remedial action designs, and remedial action work plans, the IAG

Superfund: Interagency Agreements and Improved Project... 53

establishes a review and comment process intended to result in no further comment—essentially agency agreement on the document; if either agency disagrees, it can submit the issue to dispute resolution procedures. Hence, for purposes of this chapter we consider that formal EPA approval is effectively required for these key steps. IAGs do not subject removals to formal EPA approval, although submission of certain documents is required (unless shown impracticable) before an action is taken to allow EPA to comment.[17] Removals are intended to prevent, minimize, or mitigate a release or threat of release, and are not subject to required cleanup goals, whereas a remedial action is intended to implement remedies that eliminate, reduce, or control risks to human health and the environment and generally involve establishing numerical cleanup goals. Removals do not relieve DOD of completing additional steps—such as RI/FS completion—or the full cleanup process for the site with formal EPA approval, if required to ensure long-term protection of human health and the environment.[18] In some cases, however, a removal action does fully address the threat posed by the release, and additional cleanup is not necessary.

Superfund Amendments and Reauthorization Act

In 1986, the Superfund Amendments and Reauthorization Act (SARA) added provisions to CERCLA—including section 120—specifically governing the cleanup of federal facilities. Under section 120 of CERCLA, as amended, a preliminary site assessment is to be completed by the responsible agency for each property where the agency has reported generation, storage, treatment, or disposal of hazardous waste. This preliminary assessment is reviewed by EPA, together with additional information, to determine whether the site poses a threat to human health and the environment or requires further investigation or assessment for potential proposal to the NPL.

SARA's legislative history explains that, while the law already established that federal agencies are subject to and must comply with CERCLA, the addition of section 120 provides the public, states, and EPA increased involvement and a greater role in assuring the problems of hazardous substance releases at federal facilities are dealt with by expeditious and appropriate response actions.[19] The relevant congressional conference committee report establishes that IAGs provide a mechanism for (1) EPA to independently evaluate the other federal agency's selected cleanup remedy, and (2) states and citizens to enforce federal agency cleanup obligations, memorialized in IAGs, in court.[20] Specifically, the report states that while EPA and the other federal agency share remedy selection responsibilities, EPA has the additional responsibility to make an independent determination that the selected remedial action is consistent with the NCP and is the most appropriate remedial action for the affected facility. The report also observes that IAGs are enforceable documents just as administrative cleanup orders are under RCRA and, as such, are subject to SARA's citizen suit and penalties provisions. Thus, IAGs can provide for the assessment of penalties against federal agencies for violating terms of the agreements. However, at installations without IAGs, EPA effectively has only a limited number of enforcement tools to use in compelling a recalcitrant agency to comply with CERCLA; similarly, states and citizens also lack a key mechanism to enforce CERCLA.[21]

DefenseEnvironmentalRestoration Program

Section 211 of SARA established DOD's Defense Environmental Restoration Program (DERP), providing legal authority and responsibility to DOD for cleanup activities at DOD installations and properties, including former defense sites.[22] The statute requires DOD to carry out the program subject to and consistent with CERCLA section 120. Among other things, the DERP provisions require the Secretary of Defense to take necessary actions to ensure that EPA and state authorities receive prompt notice of the discovery of a release or threatened release, the associated extent of the threat to public health and the environment, proposals to respond to such release, and initiation of any response.[23]

Executive Order 12580

Executive Order (E.O.) 12580, Superfund Implementation,[24] was issued in 1987 to respond to SARA. E.O. 12580 delegates to EPA certain regulatory authorities that the statute assigns to the President, while delegating to DOD authority for removal and remedial actions at its facilities, subject to section 120 and other provisions of CERCLA. The E.O. also constrains EPA's authorities under CERCLA section 106(a) to issue cleanup orders and under section 104(e)(5)(A) to issue compliance orders for access, entry, and inspections by the requirement that the Attorney General, DOJ concur in such actions. In practice, EPA told us it has requested DOJ concurrence approximately 15 times on unilateral section 106 orders to federal agencies and, to date, DOJ has concurred only once, when the recipient federal agency did not object.

Federal Law and Policy Affecting EPA Judicial Actions against Other Federal Agencies

CERCLA authorizes the filing of civil actions to assess and collect penalties for certain violations—such as failing to provide notice of a release—and section 120 makes each federal department subject to the full procedures and substance of CERCLA. RCRA similarly authorizes the filing of civil actions to enforce— including by assessing fines—orders issued under its imminent and substantial endangerment provision. Nonetheless, as a practical matter, court action is not an available enforcement tool to EPA against another federal agency. Federal law generally reserves the conduct of litigation in which the United States is a party exclusively to DOJ.[25] EPA officials told us the agency has not sought DOJ assistance for such actions because it is DOJ's policy that one department of the executive branch will not sue another in court.[26]

FederalFacilityComplianceAct

The Federal Facility Compliance Act of 1992,[27] which amended RCRA, authorizes EPA to initiate RCRA administrative enforcement actions against a federal agency for the cleanup of contaminated properties, among other things, as well as subjects federal agencies to RCRA's existing fines and penalties provisions. The act directs EPA to initiate administrative enforcement actions against federal agencies as it would against a private party.

History of Disputes Related to IAGs and Administrative Orders

In March 2009, we issued a report that suggested Congress may wish to consider expanding EPA's enforcement authority to give the agency more leverage to better satisfy statutory responsibilities with agencies that are unwilling to enter into IAGs where required under CERCLA.[28] The report was issued following DOD's February 2009 agreement with EPA that appeared to resolve a long dispute by determining that the 11 IAGs outstanding at the time would be completed using an IAG between the Army and EPA for Fort Eustis, Virginia, as a template. In addition, EPA agreed to rescind each administrative cleanup order upon the effective date of an installation's IAG.

Soon after this approach for resolving outstanding disputes was agreed to by EPA and DOD leadership, some progress was made in signing IAGs. For example, the Army signed an IAG for Fort Meade Army Base in June 2009. Likewise, the Air Force signed IAGs for McGuire AFB, Brandywine Defense Reutilization and Marketing Office Salvage Yard, Langley AFB, and Hanscom Field AFB, by November 2009.

In the absence of the required IAGs, DOD, at some installations, took a few actions toward cleanup and, at others, proceeded with some cleanup activities—including investigations, removals, and remedial actions—without EPA approvals, according to EPA officials.[29] To address continued challenges, EPA issued administrative cleanup orders at four DOD installations, either under EPA's RCRA authority,[30] or under EPA's Safe Drinking Water Act authority.[31] According to EPA officials, the agency took the unusual step of issuing the orders because it needed them to fulfill EPA's cleanup oversight responsibilities at the sites in the absence of IAGs. These administrative cleanup orders were issued as final in 2007 and 2008. In response, DOD challenged the validity of the administrative cleanup orders and asked DOJ to resolve certain questions in dispute between DOD and EPA over the terms of the IAGs and the circumstances under which EPA may issue administrative cleanup orders at such NPL installations. In December 2008, DOJ issued a letter upholding EPA's authority to issue administrative cleanup orders at DOD NPL installations in general, without discussing whether the facts supported these specific orders. DOJ's letter also supported including provisions in IAGs, such as the types of provisions that EPA regularly includes in its cleanup agreements with private parties, in addition to those specifically in CERCLA, while stating the opinion that DOD does not necessarily have to agree to all extra-statutory terms.[32]

After DOJ's letter, the Fort Meade Army Base recognized EPA's 2007 administrative cleanup order under RCRA and gave formal notice to EPA that the Army would comply with the order. However, at about the same time, the state of Maryland filed a lawsuit in December of 2008 against the Army "to force the Army to investigate fully and remediate soil and groundwater contamination resulting from years of mismanagement of hazardous substances, solid waste, and hazardous waste," and to enforce EPA's 2007 administrative cleanup order. In November 2009, the state voluntarily withdrew the suit after the Army, EPA, and two other federal agencies signed an IAG for Fort Meade. By the terms of the IAG, EPA withdrew the administrative cleanup order in October 2009.

In contrast, Air Force officials at Tyndall AFB and McGuire AFB did not give formal notice of intent to comply with EPA's administrative cleanup orders and never complied with the terms of the orders. For example, the Air

Force stated in a May 2008 letter to EPA regarding the Tyndall order, "the Air Force continues to challenge this Order as lacking legal and factual basis...I have directed my staff and Tyndall AFB to continue to conduct cleanup actions under [CERCLA] using our lead agency functions, authorities and responsibilities delegated to DOD."[33] The Air Force continues to assert that the IAG proposed by EPA does not match the agreed-to template, whereas EPA asserts the IAG does follow the template; both EPA and DOD officials told us the dispute over the IAG relates to the appendices listing the areas to be investigated and, if required, cleaned up. McGuire AFB's IAG was since signed and became effective December 2009, and EPA's 2008 administrative cleanup order was withdrawn. While Tyndall remains without an IAG and its administrative cleanup order is still in effect, the Air Force counsel has asserted they are continuing "substantive compliance" with the administrative cleanup order using the CERCLA process—although EPA's order specifically requires Tyndall to use the RCRA process.[34] EPA officials stated that the agency cannot on its own impose penalties or otherwise compel compliance with the administrative cleanup order at Tyndall; to do so would require concurrence from DOJ to proceed with court action against another federal agency, which is contrary to federal policy.

A summary of the current status of IAGs is provided in Table 1. In summary, seven IAGs have been signed and have become effective. There are also four installations that do not yet have signed IAGs as of June 2010. These installations have continued to lack the IAGs required by CERCLA for an extended time frame, and include three Air Force installations and one Army base: Air Force Plant 44 in Arizona, Andrews AFB in Maryland, Tyndall AFB in Florida, and Redstone Arsenal in Alabama. For summary information on the source of contamination and status of cleanup at these 11 NPL installations, see appendix II.

BECAUSE OF DIFFERING EPA AND DOD PERFORMANCE METRICS AND DOD'SFAILURE TO OBTAIN EPA APPROVALS, STATUS OF DOD CLEANUP IS UNCLEAR

Because EPA and DOD use different terminology and metrics to report investigative and remedial work at defense installations, determining the status of cleanup at Fort Meade, McGuire AFB, and Tyndall AFB is challenging. EPA's data suggest that DOD's progress at these installations was limited primarily to the early study or investigative phase, whereas DOD's data suggest that some work in the later remedial action or cleanup phase has taken place at these installations. As DOD did not obtain EPA's concurrence with some of the cleanup actions it took at these installations, it may need to conduct additional work even on reported completed actions as a result of EPA requirements.

Cleanup Work Has Begun, but EPA's and DOD's Differing Performance Metrics and Reporting Practices Result in Differing Interpretations of Progress

Twenty or more years after contamination was first reported at Fort Meade, McGuire AFB, and Tyndall AFB, EPA reports that environmental cleanup generally remains in the early, investigative phases of the CERCLA process, with little progress in achieving long-term remediation of contaminated sites at these installations. While DOD's data suggest that some remedial action work has taken place, EPA and DOD have differing interpretations of the level of cleanup achieved at these installations, in part because the agencies use different terminology and performance metrics to assess cleanup. EPA's terminology and metrics are based on the Superfund program, including some that are unique to federal facilities, while DOD's terminology and metrics are based on the DERP program, which DOD is directed to conduct in accordance with CERCLA. Specifically:

- EPA divides installations into numbered "operable units" (OU), which may represent the type of action to be taken, such as the removal of drums and tanks from the surface of an installation; the geographic boundaries of the contamination; or the medium that is contaminated, such as groundwater.
- DOD divides installations into smaller geographic areas of contamination called "sites." These sites are typically scoped narrowly to allow for targeting work on actions that can be accomplished efficiently—for example, a building or waste disposal area where a potential or actual release of hazardous substances, pollutants, or contaminants may have occurred may be considered a "site," while adjacent buildings with similar operations are considered as separate sites. DOD's sites are sometimes smaller than EPA's OUs; therefore there may be multiple DOD sites in one EPA OU.

Superfund: Interagency Agreements and Improved Project... 59

Table 1. Status of IAGs for 11 DOD NPL Installations Lacking IAGs as of February 2009

State	Installation	Status of IAG	Date added to the NPL	Date IAG signed	Effective date of IAG
Ala.	Redstone Arsenal (Army)	Under negotiation	05/31/94	[a]	[b]
Ariz.	Air Force Plant 44 Air Force Base (Tucson Int'l Airport Area)	Under negotiation	09/08/83	[c]	[b]
Fla.	Naval Air Station Whiting Field	Signed and in effect	05/31/94	03/09/09	07/10/09
Fla.	Tyndall Air Force Base	Under negotiation	04/01/97	[c]	[b]
Hawaii	Naval Computer Telecom-munication Area Administrative Master Station	Signed and in effect	05/31/94	03/24/09	07/28/09
Mass.	Hanscom Field (Air Force)	Signed and in effect	05/31/94	09/18/09	12/02/09
Md.	Fort Meade (Army)	Signed and in effect	07/28/98	06/19/09	10/06/09
Md.	Andrews Air Force Base	Under negotiation	05/10/99	[a]	[b]
Md.	Brandywine Defense Reutiliz-ation and Marketing Office Salvage Yard (Air Force)	Signed and in effect	05/10/99	11/25/09	3/30/10
N.J.	McGuire Air Force Base	Signed and in effect	10/22/99	09/15/09	12/01/09
Va.	Langley Air Force Base	Signed and in effect	05/31/94	09/30/09	12/21/09

Source: GAO analysis of EPA data.

[a] Signed by EPA; awaiting DOD signature.

[b] Not in effect.

[c] Not signed. Notes:

GAO found in a prior report that as of February 2009, these installations lacked IAGs. Since February 2009, EPA has added another DOD property, Fort Detrick Area B Ground Water in Maryland, to the NPL; see 74 Fed. Reg. 16126 (2009). See GAO, *Superfund: Greater EPA Enforcement and Reporting Are Needed to Enhance Cleanup at DOD Sites*, GAO-09-278 (Washington, D.C.: Mar. 13, 2009).

CERCLA § 120 provides that within 6 months of a federal property's listing on the NPL, the lead agency shall commence an RI/FS. 42 U.S.C. § 9620(e)(1) (2010). Then, within 180 days following EPA's review of the RI/FS report, the head of the lead department "shall enter into an interagency agreement with the Administrator for the expeditious completion by such department...of all necessary remedial action at such facility." 42 U.S.C. § 9620(e)(2) (2010). As noted previously, since the RI/FS culminates in a ROD, EPA has interpreted the IAG trigger as the first signed ROD at a federal property, but seeks IAGs as early as practicable, so as to guide all steps in the cleanup process.

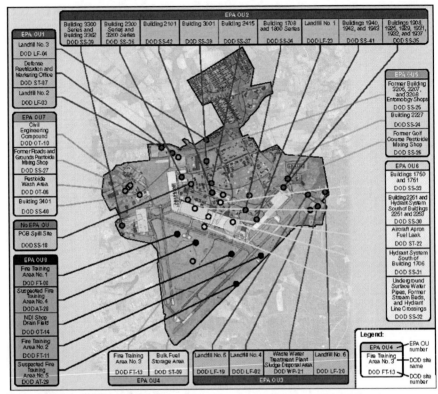

Sources: EPA, DOD, and GAO.

Notes: DOD Site Identification Codes: AT – All Training; FT – Fire Training; LF – Landfill; OT – Other; SS – Spill Site; ST – Storage; WP – Waste Pits.

EPA and DOD assign separate designations to the contaminated areas being cleaned up under CERCLA at DOD installations on the NPL. As seen in figure 1, EPA assigns names and consecutive numbers to the contaminated areas, which it refers to as "operable units" (OU). DOD delineates the same contaminated areas into smaller parts that they refer to as "sites." These sites are given both a title and a number by DOD. One EPA OU is often composed of several DOD sites. Figure 1 demonstrates the overlap and confusion caused by the various terms used to describe the same contaminated areas.

Figure 1. Map of McGuire AFB depicting EPA's and DOD's Designations of Cleanup Areas

The differing nomenclature can make it difficult to interpret and compare the information DOD reports annually to Congress with what EPA lists in its Comprehensive Environmental Response, Compensation, and Liability Information System (CERCLIS) database on the status of environmental cleanup at NPL sites.[35] For example, as seen in figure 1, EPA reports the progress of cleanup at McGuire AFB by tracking advancements achieved at the installation's 8 EPA OUs, while DOD reports progress according to advancements achieved at 36 DOD sites. EPA, as the regulator under CERCLA, must track progress made under the statute, and EPA officials said that units in program regulations must have precedence over DOD's internal system of measuring progress.

According to EPA data, most of the OUs at Fort Meade, McGuire AFB, and Tyndall AFB are in the RI/FS phase of environmental cleanup, which as seen in figure 2 occurs early in the CERCLA cleanup process. While the RI/FS phase historically has an average duration of 5.2 years for EPA OUs at federal facility sites on the NPL, many EPA OUs at these three bases have already been in the RI/FS phase for twice that long and are not yet complete. In fact only 3 of a total 37 OUs at these three installations have completed the RI/FS phase of the CERCLA process; those 3 EPA OUs are located at Fort Meade, and none of the OUs at McGuire AFB or Tyndall AFB have completed the RI/FS phase according to EPA.

DOD, on the other hand, reports that cleanup is further along at all three of these installations. For example, officials at Fort Meade said that environmental cleanup at their installation is at a very mature stage. In a 2008 report to Congress,[36] DOD reported that Fort Meade had achieved response complete at 61 percent of its 54 sites.[37] The achievement of "response complete," a DOD term, occurs either late in the CERCLA process after the remedy selected in the RI/FS phase is implemented, or at any time when DOD deems cleanup goals have been met and no further action is required at the site. As we previously reported,[38] we are concerned about the lack of clarity in DOD's use of this term to describe sites that have been administratively closed, with no physical cleanup.

In addition, EPA and DOD report dissimilar pictures of cleanup progress because each agency reports cleanup progress in a different way. For example, DOD reports on removals, which CERCLA defines as short-term and emergency actions to reduce risk, and for which EPA's formal approval is not required unless specified in an enforceable agreement. These actions are not necessarily designed to provide long-term protectiveness of human health and the environment, and sites where a removal has been conducted are still subject to the full CERCLA process, until no further action is appropriate. EPA tracks removals through its CERCLIS database, which also shows the remaining steps in the full CERCLA process; a removal may be the first response action taken, although one can occur at any time during the process.

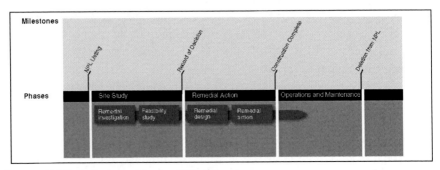

Source: GAO analysis based upon EPA data.
Note: This figure shows the general progression of steps in the NCP process under CERCLA that occur during the environmental cleanup of DOD and other sites on the NPL.

Figure 2. Environmental Cleanup Process for NPL Sites

Furthermore, EPA tracks approved cleanup actions under CERCLA that have been completed or are under way for an entire EPA OU, and records these cleanup actions by EPA OU in the CERCLIS database, where key information is made available to the public on EPA's Web site.[39] Also, EPA's current reporting system does not show cleanup progress unless the action has been achieved at all DOD sites within that OU. In contrast, DOD tracks cleanup by site through various cleanup phases as defined in the DERP, which generally aligns with CERCLA but includes additional milestones, and then reports the number of sites in each cleanup phase in its annual report to Congress.[40] For example, Tyndall AFB includes 12 EPA OUs with 12 DOD sites, with an additional 39 other sites that are not contained within any EPA OU. These additional sites are still in stages of preliminary investigation under CERCLA, according to EPA officials; DOD officials said that a number of these are regulated completely as petroleum sites under a separate program that is administered by the state of Florida, but EPA officials said they want to evaluate all of them under CERCLA, to ensure that any non-petroleum contamination that may exist is accounted for and cleaned up under CERCLA. According to EPA officials, Tyndall AFB has achieved no completed cleanup actions at the base, and it recognizes only one RI/FS action as ongoing. In contrast, DOD reported in fiscal year 2008 that Tyndall staff had completed 36 of 51 study actions for sites at Tyndall AFB, amounting to 71 percent of the study phase complete at the base.[41] The fact that DOD measures progress in smaller increments can lead to differing interpretations of cleanup. As we said earlier, DOD counts as progress the completion of each contaminated DOD site located within an EPA OU, although EPA does not count progress until action is taken at all DOD sites in that OU.[42]

In June 2009, EPA and DOD formed a working group to review and harmonize both agencies' environmental cleanup goals and metrics, with the goal of better communication between the agencies regarding cleanup progress at DOD installations on the NPL. DOD officials said they hope that the working group will minimize the inconsistencies between DOD's and EPA's goals and metrics. EPA officials said they believe the recommendations of the working group will ultimately result in fewer misunderstandings and surprises between parties that can stall cleanup actions in the future. The proposed timeline for the working group suggests the drafting of proposed recommendations in June 2011.

Status of Cleanup at These Installations Is Unclear Because DOD Did Not Obtain EPA Concurrence with Some Cleanup Actions and May Need to Do Additional Work

EPA and DOD also report very different cleanup progress at defense installations because some of DOD's reported claims of completed cleanup phases were never approved by EPA, and therefore EPA does not recognize them. In addition, where DOD has already taken actions, EPA has in some cases found that DOD's supporting documentation in the record is insufficient for EPA to approve the cleanup actions that DOD has already taken. Specifically at Tyndall, after a change in personnel at EPA, the new project manager reviewed the files and found the documentation was insufficient to support many of the previous decisions made at the base. EPA officials told us that once IAGs are in place at these installations, any unilateral cleanup actions previously taken are likely to be revisited and EPA may require work to be redone.

According to EPA officials, DOD and EPA have long agreed that, because EPA has ultimate authority under CERCLA for remedies at DOD NPL installations, EPA approval of key steps toward remedy selection is required. In practice, according to EPA officials, it is difficult for a federal facility to obtain EPA concurrence on its cleanup decisions in the absence of a signed IAG for several reasons.[43] First, from a project management perspective, EPA lacks assurance that it has had adequate involvement in key steps in the process. Second, from a compliance standpoint, EPA told us it must incorporate, among other things, an enforceable schedule and arrangements for long-term management of a remedy into a ROD, in order to approve the selected remedy at a federal facility without an IAG.[44]

At least one installation has gained EPA's concurrence with cleanup actions without an IAG through effective interagency cooperation. However, two of the three DOD installations we examined for this chapter—Tyndall AFB and Fort Meade—moved forward with cleanup actions, including remedies, without a signed IAG or ROD. For example, EPA's records for Tyndall AFB show that DOD made decisions at a number of sites without the required concurrence of EPA.

Despite the lack of IAGs, DOD submitted a variety of documents for EPA review at each of the three selected installations. However, without an IAG, there are no agreed-upon time frames for review and comment and no overall work plan to provide predictable schedules for DOD or EPA. With an IAG, EPA's typical primary document review times would be 60 days; however, DOD officials told us that EPA reviews sometimes take longer, with or without an IAG. As a result, DOD officials said that, in some cases, DOD moved forward without EPA concurrence, while in other cases DOD may have delayed planned actions. For example, EPA provided comments on a preliminary assessment of munitions sites at Tyndall AFB that included concerns about how the munitions at these sites could affect other nearby hazardous substances sites. However, EPA took approximately 4 months after receiving the assessment from DOD to submit the comments. As a result, DOD officials told us they finalized the preliminary assessment before receiving EPA's comments because they wanted to close out the contract.[45] On the other hand, without the predictable schedules provided by an IAG, EPA officials told us they could not predict the flow of documents from DOD they would have to review. EPA officials told us that DOD at times submitted few documents for review, while at other times, an overwhelming number of documents, making it difficult for EPA to allocate resources for review and comment.[46] We could not verify long-term trends in the volume of document submission and in document review times because neither EPA nor DOD maintains a consistent, verifiable, and long-term management system for tracking documents submitted or reviewed. For example, DOD said that the three installations have only maintained document tracking systems for the last 2 to 4 years.

DOD officials told us they received EPA approval of some cleanup actions in informal meetings—referred to as partnering meetings—but could not provide documentation. EPA officials noted that these meetings were never intended to replace the formal process mandated by CERCLA and that

Superfund: Interagency Agreements and Improved Project... 65

such decisions were not formally documented, as needed for EPA to approve the proposed remedy selection and as required for the administrative record. CERCLA requires the lead agency, in this case DOD, to establish an administrative record upon which DOD bases the selection of a response action. This record (1) serves as the basis for judicial review of the adequacy of the response action and (2) acts as a vehicle for public participation, since it must be made available for public inspection and comment during appropriate comment periods.[47]

A VARIETY OF OBSTACLES HAVE DELAYEDCLEANUP PROGRESS

Several obstacles have delayed cleanup at the three selected DOD installations in our review. First, the lack of IAGs has made managing installation cleanup and addressing routine matters challenging for both EPA and DOD. DOD contract management issues at some installations have affected how the work at these installations has been scoped and conducted and placed effective and efficient use of the public's resources at risk, further undermining cleanup progress. In addition, at Fort Meade Army Base, a lack of coordination with EPA and incomplete record reviews resulted in DOD personnel occupying housing at risk of contamination until they were evacuated. Further, the Air Force has failed to disclose some contamination risks at Tyndall AFB promptly, resulting in delays in taking cleanup action. We also found particular problems at Tyndall AFB, where long-standing noncompliance regarding environmental cleanup and notification has contributed to the lack of cleanup progress. Finally, EPA's ability to address noncompliance by federal facilities is limited by provisions in law, executive order, and executive branch policy.

Lack of IAGs Has Made Managing Installation Cleanup and Addressing Routine Matters Challenging, thus Delaying EPA-approved Cleanup Progress

The lack of IAGs has contributed to delays in cleanup progress at the three installations in our review. Without an IAG, EPA lacks the mechanisms to ensure that cleanup by an installation proceeds expeditiously, is properly done, and has public input, as required by CERCLA. For example, DOD officials said that EPA reviewed the proposed remedial action and provided written agreement for the Army's decision to use monitored natural attenuation—relying on natural processes to reduce the contamination in soil or groundwater without human intervention—as the remedy for groundwater contamination at the Ordnance Demolition Area at Fort Meade, which had been historically used for the demolition of unexploded munitions. However, Fort Meade did not have EPA's signature on the ROD and did not seek formal public comment. EPA officials said that additional documentation was needed to support the use of that remedy and advised Fort Meade that it was exceeding its authority. The IAG for Fort Meade provided that Fort Meade withdraw this decision document and submit a new one for EPA's review, which could result in the Army being required to carry out additional cleanup actions for that site.

Whereas an IAG would provide for negotiated deadlines designed to reflect the specific complexities at an installation, DOD's national cleanup goals may drive installations to take actions without EPA approval to meet deadlines. In particular, DOD recently set a cleanup goal for reducing risk or achieving remedy in place or response complete by 2014 for sites under DOD's Installation Restoration Program at active installations, including those at NPL-listed installations. The Air Force set an even more stringent deadline of 2012 for its sites, which Air Force officials have said is a "stretch goal" imposed to ensure that the 2014 goal is met. These deadlines were not based on evaluations of field conditions, and therefore do not necessarily reflect remaining required cleanup actions. However, DOD's use of these deadlines has acted as an incentive for DOD to proceed with actions that have not been fully vetted with EPA and the public, according to EPA officials. For example, EPA officials said that, under the pressure of the 2012 deadline, McGuire AFB has proposed monitored natural attenuation, which EPA has not approved, as a remedy for contaminated groundwater at the installation despite not having performed required analyses. EPA typically only approves monitored natural attenuation as a remedy when certain conditions exist, such as a low potential for contaminant migration and a time frame comparable to other methods of remediation. EPA said DOD did not provide evidence of these conditions to EPA, which is necessary for EPA to concur in the remedy selection, as required by CERCLA. One consequence of this gap is that the public lacks assurance that human health and the environment are adequately protected by DOD's remedy.

At installations with IAGs, the Site Management Plans include detailed schedules and become part of the IAG, thus providing a legal basis for when DOD must complete the work. Moreover, with IAGs to provide an enforceable cleanup schedule, DOD must move forward with cleanup or there will be consequences, such as penalties, for violating the terms of the agreements. These legal obligations are a key factor in DOD's sequencing of cleanup activities for funding. DOD officials told us that, in the early 1990s, the installations that had IAGs were moved to the top of the list for funding, while other installations were considered a lower priority. Also, DOD headquarters makes its funding decisions from budget requests submitted by installations; therefore, if an installation does not have an IAG and does not submit a request for funding for a particular contaminated area, DOD does not consider it in its national funding decisions.

DOD Contract Management Issues Have Undermined Cleanup Progress

DOD contracting management issues have affected how the cleanup work at the selected installations was scoped and conducted, placing effective and efficient use of the public's resources at risk, and further undermining cleanup progress. Specifically, two of the installations, Tyndall and Fort Meade, have relied extensively on performance-based contracts (PBC) to clean up installations. The third, McGuire, in 2008 awarded a PBC for 21 sites. However, PBCs can create pressure on contractors to operate within price caps and meet deadlines, which may conflict with regulatory review times and encourage DOD to take shortcuts. Both EPA and DOD officials told us that PBCs may frequently be inappropriate for some Superfund cleanup work— particularly in the investigative stages—since there can be a great deal of uncertainty in these phases. For example, initial sampling during a site investigation may lead to the need for extensive follow-up sampling that was not anticipated and therefore not provided for in the contract incentives.

While the federal government has advocated the use of PBCs in recent years for procurement of most services, federal acquisition regulations generally requiring the use of PBCs specifically exclude engineering services from this requirement.[48] DOD policy directs the services to use PBCs whenever possible—establishing the goal that PBCs be used for 50 percent of service acquisitions—but acknowledges that not all acquisitions for services can be conducted using PBCs.[49] According to federal guidelines, PBCs are not

68 United States Government Accountability Office

generally appropriate for work that involves a great deal of uncertainty concerning the parameters of the work to be performed. For example, Air Force guidance establishes the first step in using PBCs is to screen the particular project for suitability, noting that in general, a PBC may not be the right approach when the site is poorly characterized or the project would pose inordinately high risk to contractors, among other characteristics.[50] PBCs are generally better suited to work that has highly prescribed goals, such as the provision of food service or janitorial services. The general intent of PBCs is to allow contractors to determine the best way to achieve specific goals within a certain time frame for a fixed cost. When used in appropriate circumstances, PBCs can reduce costs by allowing contractors flexibility in how they provide the services.

EPA officials cited a number of problems resulting from the use of PBCs for cleanup at these three installations.[51] One problem cited is that, when PBCs are used, the contractor typically may not explore the full range of alternatives during the remedial investigation and feasibility study due to the pressure of PBC price caps to reduce the costs involved in developing these alternatives. In addition, EPA officials said, the remedies or proposals put forward by the PBCs tend to be those that do not require construction, such as monitored natural attenuation for groundwater contamination, in order to save money on the contract. For example, EPA officials said that the sole PBC contractor for 21 DOD-designated sites at McGuire AFB proposed in its contract a remedy of "no further action" for soil, sediment, and groundwater for nearly all 21 sites, along with monitored natural attenuation for groundwater at many of the sites; these approaches to address contamination at the sites were proposed prior to completing the remedial investigation, which would include a human and ecological risk assessment, feasibility study, proposed plan, public meeting, and ROD. In addition, EPA has specific guidelines on the selection of monitored natural attenuation as a remedy.

Other problems that EPA cited with using PBCs for environmental cleanup work include

- contractor's inability to carry out cleanup-related work required by EPA or other stakeholders that was not contained in the original PBC contract, such as installing monitoring wells, without contract amendment;
- unrealistic time frames for cleanup work that have not been agreed to by EPA or other stakeholders and that create an incentive for rushed work, resulting in possible rework later on;

- poor quality of documents submitted to EPA, including lack of legal review and routine failure of the installation to perform quality reviews of contractors' work, which EPA officials said were due to pressure to meet the fixed price aspect of these contracts, and which result in significant redrafting by EPA's legal staff; and
- PBC contractors—rather than DOD officials—acting as project managers to the point of decision making, rather than supporting DOD, when critical cleanup decisions require interaction between EPA and DOD officials.

In responding to a draft of this chapter, DOD noted that the department believes it has successfully used PBCs for some environmental remediation and munitions response activities. According to DOD, the PBCs include identifiable and measurable costs, schedules, and outcomes, such as acceptance by DOD and the regulatory agencies. DOD stated PBCs can benefit DOD by

- providing flexibility of scope, rather than prescriptive methods;
- allowing DOD to benefit from the expertise and emerging technologies of the private sector in solving problems during various phases of the cleanup process;
- ensuring cost control with known outcomes at the completion of the contract; and
- encouraging contractors to look for ways to reduce time and cost.

Nonetheless, Tyndall AFB officials told us that after shifting toward PBCs for cleanup work in 2004, they are now migrating away from them because there is too much uncertainty in the cleanup work needed at the base. Conversely, the Army told us that in its view, PBCs are better suited for complex work because they foster innovation from the private sector.

Poor Coordination with Regulators and Incomplete Record Reviews Resulted in DOD Personnel Occupying Base Housing at Risk of Methane Contamination until Being Evacuated at Fort Meade

At Fort Meade Army Base, a lack of coordination with EPA and incomplete record reviews led to the necessity to evacuate military personnel from housing that was at risk of methane contamination due to its construction near a dump. A contractor for Fort Meade building military housing on the base—as part of the Army's new national privatized housing construction effort—in 2003 discovered an old dump site in the area of the new housing and near an existing elementary school. Prior to construction, the Army Corps of Engineers prepared an environmental baseline survey, but it was later determined that the Corps apparently did not review key historical maps in the possession of Fort Meade indicating a former dump and incinerator in the area. The Corps, in conducting the survey, also apparently failed to use a relevant EPA report, which provided an interpretation of historical aerial photographs to identify potential hazards. According to Fort Meade documentation, once the dump was discovered, the housing contractor attempted to determine the limits of the dump and continued with construction, avoiding building directly on top of the dump site. However, according to EPA officials, Fort Meade did not involve EPA in these assessments prior to construction after the dump was discovered. Nonetheless, EPA, which had an on-site manager at the Fort Meade installation, was aware of the discovery of the dump and did not assert a role in decisions about where to locate housing. For example, EPA did not provide any written advice concerning the matter to Fort Meade. After construction was completed and the housing was occupied, methane fumes were found in 2004 below the ground in soils adjacent to the 20 houses that were built nearest the dump site and elementary school. The Army installed and operated a methane abatement system but in 2005 determined that methane was reaching the homes, and families were evacuated. These houses remain empty, and DOD is monitoring both the houses and the school for methane gas intrusion into indoor air. Thus far methane gas has not been found at an unacceptable level in the school. In addition to methane, Fort Meade has documented other contamination at the dump site, including volatile organic compounds[52] (VOC) in the groundwater, and heavy metals, polychlorinated biphenyls[53] (PCB), and VOCs in soil. Fort Meade has since prepared a preliminary assessment and site inspection[54](PA/SI) and a draft RI, which EPA has reviewed. While the Army has a policy requiring that the environmental conditions of properties be assessed,[55] it is unclear whether local Fort Meade officials were adequately involved in the preconstruction assessment, which was performed by a contractor to the Corps under the Army's national housing privatization initiative. While the Army has acknowledged that the preconstruction assessment apparently missed evidence pointing to the incinerator and dump, the Army has not explained the source of the omission—for example whether lack of adherence to policy or shortcomings in coordination and review were contributing factors. As such, it

Superfund: Interagency Agreements and Improved Project...

is unclear how the Army could prevent a recurrence of this situation in which review of key documents available to the Army may have averted construction of housing near a waste site.

Tyndall AFB's Long-standing Noncompliance Regarding Environmental Cleanup and Notification Contributed to the Lack of Cleanup Progress

Of the three installations we selected to review, only Tyndall AFB remains without an IAG. Furthermore, Tyndall has delayed cleanup progress by generally demonstrating a pattern of not complying with federal laws and regulations concerning environmental cleanup. In addition, Tyndall has on multiple occasions delayed disclosures about newly found contaminants or associated risks for months or failed to disclose them entirely, furthering delay of cleanup.

The Air Force'sFailure to Sign an IAG and Pattern of NoncompliancewithFederalLaws and RegulationsConcerningEnvironmentalCleanup Have DelayedCleanup Progress at Tyndall

After 13 years on the NPL, Tyndall AFB stands out as the only one of the three installations that received EPA administrative cleanup orders for sitewide cleanup and has not signed an IAG even though IAGs are required under CERCLA. Following DOD's issuance in February 2009 of a letter to EPA indicating its willingness to sign IAGs for the 11 installations that did not have them, most of the other installations have resolved differences with EPA and signed IAGs or are close to signing them.

As previously noted, in the absence of a signed IAG, Tyndall has delayed cleanup progress by generally demonstrating a pattern of not complying with federal laws and regulations concerning environmental cleanup under CERCLA. For example, Tyndall

- proceeded with remedies with which EPA had not concurred,
- did not seek required public input,
- failed to disclose contamination risks in a timely fashion, and
- refused to comply with the terms of the EPA-issued administrative cleanup order.

EPA officials told us DOD proceeded with cleanup remedies without EPA's written concurrence—such as signed RODs or other form of documented agreement—to protect human health and the environment, despite knowing that the work may need to be redone. Whereas the CERCLA process requires regulator oversight at federal NPL properties during cleanup activities to provide assurance of such protection, DOD officials said they relied on quarterly partnering meetings with EPA in lieu of written approvals. Tyndall has also issued contracts for work for which EPA hasn't formally concurred, potentially resulting in rework and jeopardizing public resources. For example, Tyndall authorized a PBC in June 2006 that included selecting and putting a remedy in place at a DDT-contaminated bayou within 5 years without having obtained EPA concurrence on how to proceed with the work. At an informal meeting in April 2003 that included officials from Tyndall, the Army Corps of Engineers, Fish and Wildlife Service, and the National Oceanic and Atmospheric Administration, but at which EPA officials were not present, Tyndall reportedly reached the initial decision to leave the DDT-contaminated sediment in place, with the rationale that having the DDT trapped in the sediment would be preferable to a release that could result from removing the sediment. In January 2009, Tyndall officials put forth the option to EPA officials of dredging the DDT-contaminated sediments from the bayou with the highest concentrations of contamination, proposing to carry out this ecologically sensitive and potentially risky action as a removal action for which Tyndall would not need concurrence from EPA. EPA said that a human and ecological risk assessment—which would estimate how threatening a hazardous waste site is to human health and the environment—would be needed for EPA to evaluate the proposed Air Force removal action and to determine whether it would protect the local population who catch and eat fish from the bayou. Without this information, the adequacy and protectiveness of the response action is in question.

The Air ForceFailed to Identify or DiscloseSome Contamination Risksat Tyndall in a TimelyFashion, whichDelayedCleanup Action

Tyndall AFB delayed disclosures about newly found contaminants or associated risks for months or failed to disclose them entirely. The DERP provisions of SARA require defense installations to promptly notify EPA and state regulatory agencies of the discovery of releases or threatened releases of hazardous substances, as well as the extent of the associated threat to public health and the environment.[56] However, we found that Tyndall failed to make

Superfund: Interagency Agreements and Improved Project... 73

such reports. Tyndall was also required to immediately report releases of hazardous substances to EPA according to the RCRA administrative cleanup orders,[57] but did not do so. It also did not provide potentially affected individuals with information on such releases in a timely manner, despite the requirement in CERCLA. Because Tyndall AFB failed to notify EPA of newly discovered releases, cleanup was delayed or conducted without regulatory agency oversight in recent incidents, potentially putting human health and the environment at risk.

An example of Tyndall's failure to notify EPA concerns the presence of lead—a hazardous substance under CERCLA—at the Tyndall Elementary School. Tyndall's actions have included failing to promptly report to regulators key information about the lead and its threat to public health; failing to take action to prevent children's exposure to lead shot; and potentially representing inaccurately its actions related to a cleanup, as detailed below:

- In 1992, children discovered lead shot in their playground at Tyndall Elementary School. Despite the discovery and the SARA requirement, Tyndall AFB officials did not notify EPA. Instead, Tyndall officials worked with county health officials to collect soil samples and Tyndall officials assured the public that the area was safe for children.
- From 1997 to 2000, ATSDR[58] conducted a health assessment, which was triggered by Tyndall's listing on the NPL. According to ATSDR officials, ATSDR examined Tyndall records that said the lead shot was removed and clean sand was deposited. As such, ATSDR based its assessment solely on the soil sampling results from 1992, found the contamination below levels of concern, and concurred with Tyndall taking no further action.[59] Tyndall did not conduct any follow-up surveying or sampling of the school area.
- In 2007, Tyndall issued a base-wide report—the Comprehensive Site Evaluation Phase I—that, based on a records search and visual site survey, identified inactive areas of the base where munitions, munitions constituents,[60] and unexploded munitions may have been released.[61] The report noted that Tyndall Elementary School is located on a portion of a former target range.[62]
- In 2008, Tyndall initiated the next phase of work, commencing with a site walk. Tyndall representatives observed lead shot and clay target debris on the ground surface of the playground,[63] but Tyndall did not notify EPA of this information and did not take any other action to ensure protection of the health of the children attending the school. In

March 2009, officials from the Air Force Center for Engineering and the Environment (AFCEE) visited the base and became aware of the situation and pressed Tyndall to expedite sampling that would assess potential risks. As a result, sampling of the school yard was included in the next phase of work.

- Once these samples were taken in May 2009, they showed elevated lead in the soils exceeding state standards. The base did not notify EPA until 22 days later—in contrast to the DERP statute's requirement of prompt notification of a threat, as well as the RCRA order's requirement, which states that the EPA must be notified immediately of any release of a hazardous substance.
- Once notified, EPA officials said they called for Tyndall to take appropriate action, including an emergency removal to reduce risk and notifying students' parents. Tyndall officials told us they initiated funding for a removal action before notifying EPA of sampling results and discussing the action with EPA.
- In 2009, ATSDR also became involved at the site again, and is currently conducting a health consultation. According to ATSDR officials, EPA requested the consultation in June 2009. Following the request, ATSDR notified its Air Force liaison, who then initiated the formal request on July 7, 2009.

When asked about these events, Tyndall officials stated they had always known lead shot could be there, and said they believed EPA also knew of this potential. Tyndall officials told us they did not conduct a cleanup following the 1992 discovery, although they agree that lead shot was found in the playground in 1992 and Tyndall officials subsequently assured parents that the area was safe.[64] Furthermore, Tyndall representatives disagreed with ATSDR's account that the lead shot had been removed and clean sand placed in the area – information upon which ATSDR relied in focusing its 2000 review on lead in soil exclusively and concluding the site did not pose a health hazard. In 1992, CERCLA and the DERP statute were in effect and well-established, and since lead is a CERCLA hazardous substance, DOD was legally required to conduct any response in accordance with CERCLA and its standards. Thus, Tyndall officials either left the lead shot in place with essentially no response other than to reassure parents of the schoolchildren, or conducted a response outside of CERCLA. While Tyndall officials now state that the lack of response with respect to the lead shot itself was based on its belief that ATSDR found the lead shot not to pose a health hazard, the ATSDR report was not issued until 2000 while Tyndall decided not to conduct a response action years earlier, in 1992.

Regarding Tyndall's lack of action on the discovery of lead shot, Tyndall officials did not take steps until 2009 to protect children from potential exposure, despite their statements that they knew from 1992 forward that lead shot could be present at the school, because they did not believe there were any health risks.[65] Tyndall officials further stated that they believed the ATSDR health assessment found no health risk from the lead shot. However, because ATSDR understood the lead shot had been removed, the ATSDR assessment was based solely on the soil lead levels reported by the Air Force to have been found in 1992 and did not address any subsequent risks from the presence of lead shot after 1992 (e.g., from direct contact and the possibility of increased soil levels from leaching).[66] Moreover, the ATSDR assessment had a narrow objective—to evaluate the potential human health effects associated with exposure to certain environmental conditions at several areas on the base—and was not intended as a substitute for the CERCLA process, which provides for investigations to determine whether a remedial action is required based on both human health and the environment. For example, as ATSDR focused on the likely exposure of children, it discounted certain soil samples with lead levels above its screening values because the agency determined children were unlikely to play in those areas; however, those samples are relevant for CERCLA purposes.

Finally, while Tyndall officials have not denied knowledge of the presence of lead shot in the playground prior to June 2009 (when Tyndall reported high lead levels in the soil), they were unable to identify or document when base officials or contractors became aware of the lead shot and clay target debris on the ground surface of the playground. Because Tyndall failed to promptly notify EPA of the release observed prior to March 2009, as required by the administrative cleanup order as well as the DERP provisions of SARA, EPA did not have the information needed to ensure Tyndall's actions were protective of the health of the schoolchildren.[67] Only at the urging of the Air Force Center for Engineering and the Environment did the base conduct sampling, and only when the results showed high levels of lead in soils did the base inform EPA of the lead shot. In summary, the base failed to take appropriate action to prevent lead exposure until June 2009—months after discovering the debris at the surface during the school year, when children were potentially exposed to lead in this material.[68] Figure 3 shows how visible the lead shot was on the school playground.

Tyndall's failure to disclose the lead at the schoolyard is not an isolated failure to disclose contamination risks.

- In late 2007, Tyndall discovered the Mississippi Road Landfill but delayed a year before reporting the discovery to EPA in October 2008.
- Tyndall discovered discarded smoke signal flares, which are hazardous waste under RCRA, in late October 2009 and delayed reporting this to EPA for about a month until November 2009.

EPA Is Limited in How It May Respond to Noncompliance by Federal Facilities

EPA's ability to pursue enforcement actions against federal agencies is limited by provisions of law, executive order, and executive branch policy. Specifically, EPA may issue CERCLA orders seeking information, entry, inspection, samples, or response actions from federal agencies only with DOJ's concurrence.[69] In practice, EPA told us it has requested DOJ concurrence approximately 15 times on unilateral section 106 orders to federal agencies and, to date, DOJ has concurred only once, when the recipient federal agency did not object. Moreover, under federal law, DOJ—and not EPA—is the sole representative authorized to conduct litigation on behalf of the federal government in judicial proceedings, including those arising under CERCLA. This provision, in conjunction with a long-standing DOJ policy against one federal agency suing another in court, has effectively precluded EPA judicial actions against sister federal agencies. However, EPA retains whatever enforcement provisions are contained within an IAG, such as stipulated penalties that may be established within a penalty provision in the agreement. For those installations without an IAG, EPA effectively has no enforcement tools available, without DOJ concurrence, to compel agency compliance with CERCLA.[70]

Source: EPA.

Figure 3. Lead Shot on School Playground at Tyndall Air Force Base in June 2009

CONCLUSIONS

Cleaning up the most seriously contaminated DOD installations is a daunting task, especially when these properties are in ongoing use by DOD components. We recognize that DOD's primary mission is ensuring the nation's defense, and that DOD is currently focused on ensuring its components' readiness for wars in Iraq and Afghanistan. Nonetheless, the environmental problems at the three installations addressed in this chapter have persisted for more than 20 years since laws requiring their cleanup were enacted. DOD and its components have environmental responsibilities to EPA as well as responsibilities to the public and the military personnel stationed at its installations. Despite some progress in the early investigative stages made by the installations we reviewed, we believe that DOD, the Air Force, and the Army are not fully upholding these responsibilities at the three installations.

DOD has expressed its commitment to full and sustained compliance with federal, state, and local environmental laws and regulations that protect human health and preserve natural resources. However, until the current challenges—including the lack of uniform measures for DOD and EPA to report cleanup progress, the absence of IAGs at some installations, the failure to disclose newly discovered contamination at some installations as required by provisions in SARA, and the continued disagreement over proposals for the use of monitored natural attenuation and other nonconstruction remedies, and over DOD's use of PBCs—are addressed, delays in cleaning up these three installations will likely persist.

Section 120 of CERCLA was enacted in 1986 amidst concerns that federal facilities on the NPL were taking too long to get cleaned up and contained key provisions aimed at eliminating stalemates, such as those that were occurring over IAGs. Yet, the IAGs required by law are still outstanding at several NPL installations after more than a decade of effort. While EPA is charged with regulating cleanup of federal NPL sites, without IAGs and lacking independent authority to enforce CERCLA, EPA has little leverage to facilitate compliance at such sites. While EPA ultimately issued administrative cleanup orders at these three installations under other environmental laws, the agency is nonetheless limited in its ability to enforce these orders because DOJ policy generally precludes bringing suit on behalf of one federal agency against another.

In the absence of the IAGs, EPA attempted to work with the services over the past decade by offering technical support and in many cases participating in informal meetings with DOD officials, while the services provided numerous documents to EPA. However, we believe that these interactions, while well intentioned, contributed to a less rigorous approach that interfered with the collection of documents such as formal approvals for the administrative record, and led to insufficient communication between the agencies on significant issues such as risk and approvals. Further, without the more predictable time frames as would be provided with an IAG, EPA and DOD resorted to less formal document review processes—including a lack of clarity on document review times and on whether agreements had been reached on key decisions—leading DOD to sometimes move forward in the cleanup process without EPA's concurrence. Together, these informal approaches contributed to disagreements between the agencies, further delayed cleanup, and resulted in a lack of transparency and accountability to Congress and the public.

RECOMMENDATIONS FOR EXECUTIVE ACTION

We are making six recommendations, as follows:

To provide greater assurance that cleanup progress is being measured accurately and consistently, and to build off of the existing DOD and EPA working group's initial efforts, we recommend that the Secretary of Defense and Administrator of EPA develop a plan with schedules and milestones to identify and implement a uniform method for reporting cleanup progress at the installations and allow for transparency to Congress and the public.

To ensure that outstanding CERCLA section 120 IAGs are negotiated expeditiously, should the agencies continue to be unable to execute a signed IAG within 60 days of this chapter, we recommend the Administrator of EPA pursue amendments to E.O. 12580 to (1) delegate to EPA unconditionally the independent authority to issue unilateral administrative orders under section 106(a) to executive agencies, and (2) cause the existing delegation of CERCLA remedial action authorities at NPL-listed sites to DOD to be conditional on, for example, the existence of a signed IAG or on DOD's submission of detailed monthly reports to CEQ and Congress concerning the status of IAG negotiations at such sites.

To ensure that DOD promptly reports new hazardous releases to EPA and other stakeholders (including potentially injured parties, the National Response Center, and the states), we recommend that the Secretary of Defense develop guidance for components concerning the proper notification when a new release is discovered or significant new information about a previously known release is obtained. The guidance should at a minimum address timing and contents of such notice, as well as meet the requirements of CERCLA § 103(a) and 111(g) and 10 U.S.C. § 2705(a).[71]

To improve project management at DOD NPL sites regarding the use of contractors, we recommend that the Secretary of Defense ensure that the services make a determination of appropriateness, using Office of Management and Budget criteria and service guidance, before using PBCs for Superfund cleanup.

To ensure that DOD NPL sites utilize monitored natural attenuation as the sole remedy at contaminated sites only when it is documented to meet remediation objectives that are protective of human health and the environment, we recommend that the Secretary of Defense direct the services to document compliance with relevant EPA guidance when selecting monitored natural attenuation.

80 United States Government Accountability Office

To ensure that the document review process is used effectively and to facilitate oversight and transparency between DOD and EPA, even where there are no IAGs in effect, we recommend that the Administrator of EPA establish a record-keeping system for DOD NPL sites, consistent across all regions, to accurately track documents submitted for review, including the status of approvals.

MATTER FOR CONGRESSIONAL CONSIDERATION

While EPA is charged with regulating cleanup of federal NPL sites, it has little leverage to facilitate compliance at such sites. Specifically, when a federal agency refuses to enter an IAG at an NPL site or to comply with an administrative cleanup order issued pursuant to RCRA's imminent hazard provision, EPA cannot take steps to enforce the law, such as initiating a court action to assess fines, as it would do in the case of a private party. As we suggested in 2009,[72] Congress may want to consider amending section 120 of CERCLA to authorize EPA—after an appropriate notification period—to administratively impose penalties to enforce cleanup requirements at federal facilities. This review provides further reason to emphasize such authorities to facilitate more timely and efficient compliance at federal facilities.

APPENDIX I. OBJECTIVES, SCOPE, AND METHODOLOGY

We were asked to determine (1) the status of Department of Defense (DOD) cleanup of hazardous substances at selected DOD installations subject to administrative orders and (2) obstacles, if any, to progress in cleanup at these selected sites and the causes of such obstacles.

To select installations for more detailed study from the 11 installations that were out of compliance with the Comprehensive Environmental Response, Compensation, and Liability Information System (CERCLA) in February 2009 because they did not have interagency agreements (IAG), we reviewed the 4 that were issued additional Environmental Protection Agency (EPA) cleanup orders under the Resource Conservation and Recovery Act (RCRA) or under the Safe Drinking Water Act (SDWA). These 4 installations are Air Force Plant 44 in Arizona, Fort Meade Army Base in Maryland, McGuire Air Force Base (AFB) in New Jersey, and Tyndall AFB in Florida.

EPA and DOD agreed that one of these—Air Force Plant 44, the only 1 of the 4 installations that was issued the SDWA order—was near cleanup completion and we therefore eliminated it from our selection of installations.

To determine the status of DOD cleanup of hazardous substances at the three selected installations, we toured the three installations; interviewed officials from DOD, EPA, DOD contractors, and the Public Employees for Environmental Responsibility, a public interest group; and attended an installation's Restoration Advisory Board meeting. We reviewed numerous laws, guidance, and technical documents, including CERCLA, RCRA, DOD Defense Environmental Restoration Program (DERP) guidance and annual reports to Congress, decision documents, and correspondence between EPA and DOD. We reviewed and analyzed information on cleanup progress from EPA's Comprehensive Environmental Response, Compensation, and Liability Information System (CERCLIS) information system, the three EPA regions that monitor cleanup at the installations, and from the individual DOD installations.

To identify any obstacles to progress in cleanup at the selected installations and the causes of such obstacles, we interviewed officials from DOD, EPA, the Agency for Toxic Substances Disease Registry (ATSDR), the Fish and Wildlife Service, and the Architect of the Capitol, as well as state officials from Florida, Maryland, and New Jersey, and the Public Employees for Environmental Responsibility. We reviewed numerous laws, guidance, orders, and technical documents, including EPA guidance on the appropriate selection of cleanup remedies; decision documents; correspondence between EPA and DOD; internal EPA and DOD documents; ATSDR reports; federal contracting guidelines; and GAO reports on government contracting and project management.

We conducted this performance audit from January 2009 to July 2010 in accordance with generally accepted government auditing standards. Those standards require that we plan and perform the audit to obtain sufficient, appropriate evidence to provide a reasonable basis for our findings and conclusions based on our audit objectives. We believe that the evidence obtained provides a reasonable basis for our findings and conclusions based on our audit objectives.

Appendix II. Cleanup Progress (According to EPA) at DOD Sites LackingIAGs in Early 2009

In February 2009 DOD sent EPA an e-mail indicating its renewed willingness to accept the Fort Eustis Federal Facility Agreement as the model for DOD's remaining site agreements under CERCLA. At that time EPA reported there were 12 DOD installations on the National Priorities List (NPL) without agreed-upon IAGs, as required under CERCLA. (Since then, DOE and EPA acknowledge there are only 11 installations without IAGs for which DOD is responsible. They exclude the Middlesex Sampling Plant, which is the responsibility of the Army Corps of Engineers.) For a detailed list of the 11 DOD installations, see table 2.

EPA told us that since February 2009, progress has been made and IAGs were signed and made effective for Fort Meade in Maryland, Naval Computer and Telecommunications Area Master Station in Hawaii, and Whiting Field in Florida. In addition, as of June 2010 the remaining four installations that lack signed IAGs include Andrews AFB in Maryland, Tyndall AFB in Florida, Redstone Arsenal in Alabama, and Air Force Plant 44 in Arizona.

Background on Installation

The Fort Meade Army Installation is located approximately halfway between Baltimore, Maryland, and Washington, D.C., near Odenton, Maryland, and has been a permanent United States Army Installation since 1917. Fort Meade once occupied approximately 13,500 acres of land, but currently occupies approximately 5,142 acres after parcels of land were transferred to the U.S. Department of the Interior, the U.S. Architect of the Capitol, and Anne Arundel County, Maryland. Fort Meade's mission is to provide base operations support for activities of over 80 partner organizations from all four Department of Defense (DOD) military services and several federal agencies. Some of the major tenant agencies include the National Security Agency, the Defense Information School, the U.S. Army Intelligence and Security Command, the Naval Security Group Activity, the 70th Intelligence Wing (Air Force), the 902nd Military Intelligence Group (Army), and the U.S. Environmental Protection Agency (EPA).

Table 2. IAG Status and Other Details for 11 DOD Installations on the NPL that Lacked IAGs as of February 2009

Installation name and state	Discovery date	Final listing on the NPL	IAG status	EPA operable units	DOD sites[a]	Completed cleanup progress installation-wide	Ongoing cleanup progress installation-wide	Examples of known contaminants
Andrews Air Force Base (Md.)	6/1/1981	5/10/1999	Signatures expected soon	14	29	7 RI/FS actions, 7 RODs, 3 remedial designs, 3 remedial actions	7 RI/FS actions, 1 remedial design, and 1 remedial action	Lead, mercury, chromium, cadmium, VOCs, semi-VOCs, polynuclear aromatic hydrocarbons, and PCBs
Brandywine Defense Reutilization and Marketing Office (DRMO) (Md.)	7/24/1991	5/10/1999	Signed and effective	3	3	1 removal, 1 RI/FS, 1 ROD, 1 remedial design	1 removal, 1 remedial action	PCBs, semi-VOCs, VOCs, PCE, TCE, and iron
Fort George G. Meade (Md.)	12/1/1979	7/28/1998	Signed and effective	17	54	7 removals, 3 RI/FS actions, 3 RODs, 1 remedial design	13 RI/FS actions	VOCs, pesticides, explosive compounds, PCE, TCE, and pesticides
Hanscom Field/Hanscom Air Force Base (Mass.)	6/1/1981	5/31/1994	Signed and effective	2	22	Construction complete		Chlorinated solvents, jet fuel, , PCBs, VOCs, and other petroleum compounds
Langley Air Force Base/NASA Langley Research Center (Va.)	10/17/1989	5/31/1994	Signed and effective	29	66	4 removals, 16 RI/FS actions, 18 RODs, 18	1 removal, 4 RI/FS actions, 1 remedial	PCBs, PCTs, photofinishing

Table 2. (Continued)

Installation name and state	Discovery date	Final listing on the NPL	IAG status	EPA operable units	DOD sites[a]	Completed cleanup progress installation-wide	Ongoing cleanup progress installation-wide	Examples of known contaminants
						remedial designs, 9 remedial actions	design, 7 remedial actions	wastes, solvents, lubricating oils,hydraulic fluids, mercury, and pesticides
McGuire Air Force Base (N.J.)	11/1/1974	10/22/1999	Signed and effective	8	36	4 removals	8 RI actions	VOCs, PCBs, inorganic hazardous substances, nickel, and mercury
Naval Computer and Telecommunications Area Master Station Eastern Pacific (Hawaii)	5/1/1987	5/31/1994	Signed and effective	5	30	6 RI/FS, 2 RODs, 2 remedial designs, 2 remedial actions	2 RI/FS. 4 RODs	PCBs, creosote, mercury, chlorinated and nonchlorinated solvents, hydra-ulic fluid, paint thinners, and TCE
Tucson International Airport Area of Air Force Plant #44 (Ariz.)	12/1/1979	9/8/1983	Not signed, in negotiation	2	13	7 removals, 2 RI/FS actions, 4 RODs, 2 reme-dial designs, 5 remedial actions	1 remedial action	TCE, chromium, arsenic, chloroform, lead, PCBs, and VOCs
Tyndall Air Force Base (Fla.)	2/12/1988	4/1/1997	Not signed, in negotiation	12	51	In dispute	In dispute	DDT, TCE, lead, arsenic, chromium, munitions constituents, and jet fuels

APPENDIX III. PROFILE OF FORT G. MEADE ARMY INSTALLATION IN MARYLAND/EPAREGION 3

Installation name and state	Discovery date	Final listing on the NPL	IAG status	EPA operable units	DOD sites[a]	Completed cleanup progress installation-wide	Ongoing cleanup progress installation-wide	Examples of known contaminants
US ARMY/NASA Redstone Arsenal (Ala.)	11/16/1988	5/31/1994	Not signed, in negotiation	17	271	6 removals, 12 RI/FS actions, 11 RODs, 1 remedial design, 1 remedial action	27 RI/FS actions, 2 remedial actions	DDT, arsenic, mercury, perchlorate, and TCE
Whiting Field Naval Station (Fla.)	2/12/1988	5/31/1994	Signed and effective	27	47	5 removals, 22 RI/FS actions, 22 RODs, 17 remedial actions	3 RI/FS actions	TCE, arsenic, barium, copper, lead, mercury, waste solvents, fuels, and machine fluids

Source: EPA.

Note: DDT = dichlorodiphenyltrichloroethane; PCB = polychlorinated biphenyls; PCE = tetrachloroethylene; RI/FS = remedial investigation and feasibility study; ROD = record of decision; TCE = trichloroethylene; VOC = volatile organic compound.

[a] Number of sites is as of the end of 2008.

NPL Listing History and Known Contaminants

The EPA placed Fort Meade on the National Priority List (NPL) on July 28, 1998, after an evaluation of contamination due to past storage and disposal of hazardous substances at the Defense Reutilization and Marketing Office, Closed Sanitary Landfill, Clean Fill Dump, and Post Laundry Facility. Contamination at these sites included solvents, pesticides, polychlorinated biphenyls (PCB), heavy metals, waste fuels, and waste oils. Moreover, elevated levels of volatile organic compounds (VOC), pesticides, and explosives compounds have been detected in underlying aquifers and low levels of VOCs, including tetrachloroethylene (PCE) and trichloroethylene (TCE), and pesticides have been detected in residential wells located off-base in Odenton, Maryland.

Issuance of RCRA 7003 Order

On August 27, 2007, EPA issued a unilateral Administrative Order under the Resource Conservation and Recovery Act (RCRA) section 7003 for Fort Meade under its authority to address solid and hazardous wastes that may present an imminent and substantial endangerment to health or the environment. The RCRA Order requires the Army to assess the nature and extent of contamination, determine appropriate corrective measures, and implement those measures. The Order was motivated by the absence of a signed interagency agreement (IAG) between EPA and DOD, as required by section 120 of CERCLA, and which would establish a framework for EPA's involvement. EPA and the Army could not come to an agreement on the IAG due to several issues. For many years, the Army maintained the position that since EPA took only four sites into consideration for listing Fort Meade on the NPL, it would negotiate an IAG for only those four sites.[73] EPA's position on the other hand has been that the 14 Areas of Concern on the Fort Meade property and 3 Areas of Concern on the adjacent transferred property should be included in the language of the IAG. Another major disagreement centers on groundwater contamination issues at the base, a common problem on DOD installations. The RCRA Order consequently required the Army to move forward with cleanup of all these hazardous waste sites. Fort Meade officials accepted the order in December 2008.

Recent Developments in IAG Negotiation

While as of March 2009, Fort Meade was out of compliance with the RCRA Order, in June of 2009, DOD and EPA reached an agreement and an IAG for Fort Meade was signed by all parties.[74] The IAG became effective in October of 2009, after the required public comment period. Per the terms of the IAG, the EPA has rescinded the RCRA Order at Fort Meade.

APPENDIX IV. PROFILE OF MCGUIRE AIR FORCE BASE IN NEW JERSEY/EPA REGION 2

Background on Installation

McGuire Air Force Base (AFB) is located in south-central New Jersey near the town of Wrightstown, which is approximately 20 miles southeast of Trenton, and occupies about 3,536 acres within the boundaries of the Pinelands National Reserve. McGuire AFB began operations in 1937 functioning under the control of the U.S. Army until 1948 when the facility's jurisdiction was transferred to the Air Force. McGuire AFB is home to five units of command, including the 87th Air Base Wing (the host wing), 108th Air Refueling Wing, 305th Air Mobility Wing, 514th Air Mobility Wing, and 621st Contingency Response Wing. McGuire AFB's mission is to provide joint installation support for McGuire AFB, Fort Dix (Army), and the Naval Air Engineering Station Lakehurst. McGuire AFB is the Department of Defense's (DOD) first and only joint base to consolidate Air Force, Army, and Navy installations. The base provides airlift capabilities to place military forces into combat situations.

NPL Listing History and Known Contaminants

The Environmental Protection Agency (EPA) placed McGuire AFB on the National Priorities List (NPL) on October 22, 1999. The initial sites responsible for McGuire AFB's inclusion on the NPL include: (1) Zone 1 Landfills (comprised of Landfill Nos. 4, 5, and 6; (2) Landfill No. 2; (3) Landfill No. 3; and (4) the Defense Reutilization and Marketing Office. Examples of contaminants found on McGuire AFB sites include volatile organic compounds; polychlorinated biphenyls; trichloroethylene; semivolatile organic compounds; polycyclic aromatic hydrocarbons; total petroleum hydrocarbons; pesticides; and metals, such as nickel and mercury. There are 42 contamination sites[75] in total at McGuire AFB, where 36 sites are located on the base and 6 sites, which are not included in McGuire AFB's NPL listing, are located at the Boeing Michigan Aeronautical Research Center Missile Facility. According to McGuire AFB officials, the sites that have the greatest priority for cleanup include the landfill sites, which were responsible for McGuire AFB's listing on the NPL, the Bulk Fuel Storage Area, the Triangle area, the Defense Reutilization and Marketing Office site, the C-17 Hangar site, the Fuel Hydrant Area, and the Pesticide Shop Area.

Issuance of RCRA 7003 Order

On July 13, 2007, EPA issued a RCRA Administrative Order under section 7003 for McGuire AFB, which became effective on November 26, 2007. EPA issued the order under its RCRA authority to address solid and hazardous wastes that may present an imminent and substantial endangerment to health or the environment. The RCRA Order requires McGuire AFB to assess the nature and extent of contamination, determine appropriate corrective measures, and implement those measures. The Order was motivated by the absence of an IAG between EPA and DOD at McGuire AFB, according to EPA officials.

Recent Developments in IAG Negotiation

On December 7, 2007, the Air Force notified EPA by letter that it considered the RCRA Order for McGuire AFB to be invalid. The Air Force officials said that the contamination sites listed in the Order, which were also included in a draft IAG for the base, had not been updated since 2001. According to EPA, the RCRA Order was based on site information from McGuire AFB's outdated documents, since those were the only sources of the information available to EPA at the time. In addition, the officials at McGuire AFB believed that EPA's issuance of the RCRA Order was politically motivated and that it slowed cleanup progress at the base. For example, they believed that EPA did not approve McGuire AFB's site management plan

Superfund: Interagency Agreements and Improved Project...   89

(SMP)— related to cleanups under the Comprehensive Environmental Response, Compensation, and Liability Act (CERCLA)—because the RCRA Order was in place. However, prior to the issuance of the RCRA Order, McGuire AFB had not submitted an SMP and only provided EPA with individual fact sheets for contamination sites on the base. McGuire AFB submitted a revised draft SMP in July 2009. Officials from the Air Force said that the Air Force would continue to exercise its CERCLA responsibilities at McGuire AFB to accomplish the substantive cleanup work that EPA sought to impose in the RCRA Order. However, this did not stop EPA's involvement with the cleanup activities at McGuire AFB, as EPA continued to work with Air Force officials on the RCRA Facility Investigation phase at McGuire AFB. According to EPA officials, McGuire AFB was not in compliance with the RCRA Order as it had not complied with deadlines set forth in the Order and refused to follow the outlined cleanup process. It is EPA's opinion that only after EPA's issuance of the RCRA Order did McGuire AFB begin submitting the required documentation. However, McGuire AFB overwhelmed EPA's document review process by submitting the required documents all at once. Following DOJ's letter upholding EPA authority to issue the RCRA Order, as a matter of law, DOD asserted that fulfilling CERCLA requirements fulfilled the Order's RCRA requirements. Nonetheless, progress was made on the IAG negotiations at McGuire AFB. In October 2009, an IAG was signed by all the appropriate parties for McGuire AFB and it became effective on December 1, 2009, following a public comment period.

APPENDIX V. PROFILE OF TYNDALL AIR FORCE BASE IN FLORIDA/EPAREGION 4

Background on Installation

Tyndall Air Force Base (AFB) occupies approximately 29,000 acres on a peninsula near Panama City, Florida. The base was initially activated in 1941 as a gunnery school for the Army Air Corps, then as an air tactical training school in 1946, and finally designated as an Air Force base in 1947. Currently, Tyndall AFB contains the 325th Fighter Wing, which has a mission of pilot and maintenance training for the F-15 Eagle and F-22 Raptor fighter jet squadrons, weapons system controllers training, and the 601st Air Operations

90   United States Government Accountability Office

Center activities. Tyndall AFB is also part of the Air Education and Training Center.

NPL Listing History and Known Contaminants

The Environmental Protection Agency (EPA) placed Tyndall AFB on the National Priorities List (NPL) on April 1, 1997, primarily due to DDT contamination in the sediment of Shoal Point Bayou. Shoal Point Bayou is a tidal creek used as a waterway for barges and small ships to deliver petroleum, oil, lubricant products, and building supplies to the base. In October 1985, the U.S. Fish and Wildlife Service conducted sediment sampling throughout St. Andrew Bay, including Shoal Point Bayou, and found the presence of DDT and DDT metabolites. Then in 1990, the same contaminants were detected in fish, soil, and sediment throughout the Bayou. After multiple investigations, a remedial investigation (RI) was completed for this site in 2002 by the Department of Defense (DOD); however, EPA later deemed the investigation insufficient. Additional investigations have been completed, which found higher concentrations of DDT and metabolites than previously determined. However EPA officials report that the new information on the contamination at Shoal Point Bayou was never integrated into the previous RI findings. Other areas of contamination at Tyndall AFB include the flight line and aircraft maintenance areas, oil/water separators, landfills, fire training pits, petroleum release sites, and munitions testing, disposal, and burial areas. The other contaminants of concern in soil, sediment, surface water, and groundwater at Tyndall AFB include petroleum, DDT, chlordane, TCE, vinyl chloride, pesticides, lead, benzene, arsenic, chromium, barium, and munitions constituents. DOD officials claim that Tyndall AFB currently has 16 active contamination sites after beginning its Installation Restoration Program with 39 sites.

Tyndall AFB has many cleanup challenges due to its geography and topography, which cover approximately 110 miles of coastal shoreline with a maximum elevation of less than 30 feet above mean sea level. In addition, approximately 40 percent of the land on Tyndall AFB is wetlands and there are three underlying groundwater aquifers on the base. Tyndall AFB is proceeding at many of the sites by employing a cleanup remedy of natural attenuation. One challenge is that the groundwater at the installation is highly susceptible to contamination and is used as a drinking water source on base. Another challenge is protecting Tyndall AFB's extensive wetlands and bayous, which includes protecting over 40 species of threatened and endangered plant and animal species. Finally, it is a challenge to control civilian, military, visitor, and trespasser access to areas of contamination on the base. For example, Tyndall AFB has over 110 miles of uncontrolled shoreline where recreational boaters and trespassers may gain access and be exposed to contamination. Furthermore, military and civilian workers may access areas of contamination throughout Tyndall AFB because the installation does not have a land use controls program or physical barriers, such as fences, to prevent unacceptable exposures.

Issuance of RCRA 7003 Order

Tyndall AFB cleanup and remedial investigation activities have continued in the absence of a signed IAG and without EPA concurrence. On November 21, 2007, EPA issued an Administrative Order under RCRA section 7003 for Tyndall AFB to provide EPA with an instrument to enforce cleanup and which EPA hoped would lead to a signed IAG. EPA issued the Order, which was finalized in May 2008, under its Resource Conservation and Recovery Act (RCRA) authority to address solid and hazardous wastes that may present an imminent and substantial endangerment to health or the environment. The RCRA Order requires Tyndall AFB to assess the nature and extent of contamination, determine appropriate corrective measures, and implement those measures. Tyndall AFB has maintained progress schedules for individual sites, but EPA officials say that Tyndall AFB has not submitted an integrated site cleanup schedule as part of a larger site management plan (SMP) for the entire base.

Recent Developments in IAG Negotiation

EPA officials stated that outside of their goal to reach an agreed-upon IAG, one of their other priorities is to get Tyndall AFB to submit a draft SMP in the near future. Tyndall AFB submitted one in the past, but according to EPA officials it was deficient, lacked integrated schedules, and only addressed approximately 30 contaminants on the base. However, according to EPA, Tyndall AFB is currently out of compliance with the deadlines and scope of work requirements as defined in the RCRA Order. In addition EPA officials said the Air Force has denied the Order's legitimacy by calling it a "potential Order." As of June 2010, Tyndall AFB still does not have a signed IAG.

APPENDIX VI. COMMENTS FROM THE ENVIRONMENTAL PROTECTION AGENCY

UNITED STATES ENVIRONMENTAL PROTECTION AGENCY
WASHINGTON, D.C. 20460

JUN 23 2010

Mr. John B. Stephenson
Director
Natural Resources and Environment
U.S. Government Accountability Office
Washington, DC 20548

Re: EPA comments on June 2010 Draft GAO Report titled, *Interagency Agreements and Improved Project Management Needed to Achieve Cleanup Progress at Key Defense Installations* GAO-10-348.

Dear Mr. Stephenson:

Thank you for the opportunity to review GAO's draft report entitled *Interagency Agreements and Improved Project Management Needed to Achieve Cleanup Progress at Key Defense Installations* (GAO)-10-348. GAO reviewed the status of Department of Defense (DoD) cleanup of hazardous substances at three installations – McGuire Air Force Base in New Jersey, Tyndall Air Force Base in Florida, and Fort George G. Meade in Maryland – which, at the time of your study, lacked interagency agreements required under the Comprehensive Environmental Response, Compensation and Liability Act (CERCLA).

GAO observed that Interagency Agreements (IAGs) are now in place at McGuire Air Force Base and Fort Meade, but that Tyndall Air Force Base "has delayed cleanup progress by generally demonstrating a pattern of not complying with federal laws and regulations concerning environmental cleanup" (draft, pp. 28-29). In addition, Tyndall officials "...delayed disclosures about newly found contaminants or associated risks for months or failed to disclose them entirely" (draft, p. 30). These findings are consistent with EPA's experience at this site, which remains out of compliance with a 2007 imminent and substantial endangerment order under the Resource Conservation and Recovery Act. GAO went on to observe that "...[i]n the absence of IAGs, EPA attempted to work with the services over the past decade by offering technical support and in many cases participating in informal meetings with DOD officials" that "while well intentioned, contributed to a less rigorous approach" (draft, p. 37). GAO concluded that "these informal approaches contributed to disagreements between the agencies, further delayed cleanup and resulted in a lack of transparency and accountability to Congress and the public" (draft, pp. 37-38).

Superfund: Interagency Agreements and Improved Project... 93

In light of these and other observations, GAO made three recommendations to EPA (draft, pages 38-39):

1) *"To provide greater assurance that cleanup progress is being measured accurately and consistently, and to build off of the existing DOD and EPA working group's initial efforts, we recommend that the Secretary of Defense and Administrator of EPA develop a plan with schedules and milestones to identify and implement a uniform method for reporting cleanup progress at the installations and allow for transparency to the Congress and the public."*

EPA agrees with this recommendation, and we are pursuing it through the goal harmonization project supported by DOD and EPA. Schedules and milestones for the EPA/DoD Goal Harmonization Workgroup could provide stronger cross agency support, collaboration toward performance results, and greater transparency in setting goals for cleanup milestones.

2) *"To ensure that outstanding CERCLA Section 120 interagency agreements are negotiated expeditiously, should the agencies continue to be unable to execute a signed IAG within 60 days of this report, we recommend the Administrator of EPA, pursue amendments to Executive Order 12580 to (1) condition the delegation of CERCLA authorities to DOD for its NPL-listed sites on the existence of a signed IAG, and (2) delegate to EPA unconditionally the independent authority to issue unilateral administrative orders under section 106(a) to executive agencies."*

EPA agrees that providing EPA with independent order authority under CERCLA Section 106 would strengthen EPA's ability to take appropriate enforcement against federal agencies consistent with the law's direction to apply the statute "in the same manner and to the same extent" to federal entities.

3) *"To ensure that the document review process is used effectively and to facilitate oversight and transparency between DOD and EPA, even where there are no IAGs in effect, we recommend that the Administrator of EPA establish a record-keeping system for DOD NPL sites, consistent across all regions, to accurately track documents submitted for review, including the status of approvals."*

We agree that to promote proper oversight and ensure transparency, EPA needs an improved record keeping system, particularly for the status of approvals. EPA will examine a range of implementation options for accomplishing this goal.

Again, thank you for the opportunity to comment. Please contact me if I can be of assistance, or your staff may call Bobbie Trent in EPA's Office of the Chief Financial Officer at 202 566-0983.

Sincerely,

Mathy Stanislaus
Assistant Administrator
Office of Solid Waste and
Emergency Response

Cynthia Giles
Assistant Administrator
Office of Enforcement and
Compliance Assurance

APPENDIX VII. COMMENTS FROM THE DEPARTMENT OF DEFENSE

Note: GAO comments supplementing those in the report text appear at the end of this appendix.

OFFICE OF THE UNDER SECRETARY OF DEFENSE
3000 DEFENSE PENTAGON
WASHINGTON, DC 20301-3000

ACQUISITION,
TECHNOLOGY
AND LOGISTICS

Mr. John B. Stephenson
Director, Natural Resources and Environment
U.S. Government Accountability Office (GAO)
441 G Street, N.W.
Washington, D.C. 20548

Dear Mr. Stephenson:

    This is the Department of Defense (DoD) response to the GAO Draft Report, GAO-10-348, "SUPERFUND: Interagency Agreements and Improved Project Management Needed to Achieve Cleanup Progress at Key Defense Installations," dated June 2010 (GAO Code 361033). Our detailed responses to the GAO recommendations and the matter for Congressional consideration are provided in enclosure 1. Enclosure 2 summarizes additional substantive issues that DoD has with the GAO draft report, and enclosure 3 provides our technical comments.

    The Department concurs with GAO's recommendations to the Secretary of Defense. We are committed to signing negotiated Federal Facilities Agreements (FFAs) at all 141 of our facilities listed on the National Priorities List (NPL). We have signed FFAs at 136 facilities to date, and we are actively negotiating with the Environmental Protection Agency (EPA) Regions to sign the remaining five FFAs. Following the DoD-EPA agreement to use the Fort Eustis template FFA, we signed seven of the 11 agreements that were more than 10 years overdue, and we continue to work to get agreement on the remaining four plus another more recent agreement.

    Your report raises several good points, and we have already implemented solutions to many of the problems it highlights. (The report identifies some actions taken by Tyndall Air Force Base (AFB) that we are going to investigate and assess the need for further action.) Additionally, the report points out that we have proceeded with some cleanup actions even without EPA approval. We have generally done this either because the actions did not require EPA approval or because EPA did not provide comments within a reasonable timeframe and we felt it was critical for us to proceed in order to protect human health and the environment. Moreover, some of the issues raised in the report are specific to Tyndall AFB and Fort Meade and are not representative of all DoD installations.

Superfund: Interagency Agreements and Improved Project...

The Department does not concur with GAO's suggestion that Congress consider amending section 120 of Comprehensive Environmental Response, Compensation, and Liability Act (CERCLA) to authorize EPA to administratively impose penalties to enforce cleanup requirements at federal facilities without a negotiated CERCLA interagency agreement. EPA has enforcement tools under existing environmental statutes, such as CERCLA section 109, the Resource Conservation and Recovery Act, and the Safe Drinking Water Act. Moreover, DoD has made significant progress in the last year. Thus, congressional action is not necessary, in our view.

Thank you for the opportunity to provide the Department's views.

Sincerely,

Dorothy Robyn
Deputy Under Secretary of Defense
(Installations and Environment)

Enclosures:
As stated

See comment 1.
See comment 1.

96   United States Government Accountability Office

ENCLOSURE 1
GAO DRAFT REPORT DATED JUNE 2010
GAO-10-348 (GAO CODE 361033)

"SUPERFUND: Interagency Agreements and Improved Project Management Needed to Achieve Cleanup Progress at Key Defense Installations"

DEPARTMENT OF DEFENSE COMMENTS TO THE GAO RECOMMENDATIONS

RECOMMENDATION 1: The GAO recommends that the Secretary of Defense, and the Administrator of the Environmental Protection Agency, develop a plan with schedules and milestones to identify and implement a uniform method for reporting cleanup progress at the installations and allow for transparency to the Congress and the public. (See page 38/GAO Draft Report.)

DoD RESPONSE: DoD concurs. DoD agrees that it is vitally important to track cleanup progress at our installations and to make that information available to Congress and the public in a manner that is transparent and easily understandable. DoD acknowledges that this is a challenge because DoD and EPA currently use different terms and metrics to report progress. For example, DoD tracks progress at discreet areas known as sites, while EPA tracks progress at operable units (OUs).

That is why in June 2009, DoD began actively working with EPA through a federal working group to improve communication and better correlate reporting of cleanup progress using existing performance measures at NPL installations. If the working group decides that a common metric is essential, the measure must meet certain criteria, including the evaluation of progress at DoD's site level rather than at OUs. In the late 1980's, DoD considered measuring progress at the OU level. Based on a pilot test, DoD decided measuring progress at the site level provided the fidelity and precision that DoD required to most efficiently manage cleanup.

RECOMMENDATION 2: The GAO recommends that the Administrator of EPA pursue amendments to Executive Order 12580 to (1) condition the delegation of CERCLA authorities to DoD for its NPL-listed sites on the existence of a signed interagency agreement (IAG), and (2) delegate to EPA unconditionally the independent authority to issue unilateral administrative orders under section 106(a) to executive agencies.

See comment 2.
See comment 3.
See comment 3.

Superfund: Interagency Agreements and Improved Project... 97

DoD RESPONSE: DoD nonconcurs. DoD disagrees that Executive Order 12580 should be amended to address CERCLA IAGs, especially since 136 of 141 DoD-EPA IAGs have been finalized and only 5 remain in negotiation. The signed IAGs represent great progress. It would be inappropriate to characterize the entire IAG process as flawed based on the 5 remaining agreements under negotiation, which address more complex cleanup issues. Additionally, keeping lead agency authority with DoD allows DoD to continue executing cleanup actions pending resolution of any IAG issues, which supports protection of human health and the environment. DoD looks forward to working with EPA to sign the remaining 5 IAGs using the agreed on Fort Eustis template, as GAO references on pages 10 and 44 of their draft report.

RECOMMENDATION 3: The GAO recommends that the Secretary of Defense develop guidance for components concerning the proper notification when a new release is discovered or significant new information about a previously known release is obtained. The guidance should at a minimum address timing and contents of such notice, as well as meet the requirements of CERCLA 103(a) and 10 U.S.C. 2705(a). (See page 38/GAO Draft Report.)

DoD RESPONSE: DoD concurs. DoD agrees that the identified statutes provide mandatory notification and that proper notification of new releases that exceed statutory limits and significant new information about a previously known release is necessary. That is why DoD has issued the following guidance on the subject:

- DoD Instruction entitled *Environmental Compliance*, dated April 1996 (GAO extracted the document from the DENIX web site)
- *DoD Safe Drinking Water Act Compliance Guidance*, dated September 1999
- *Management Guidance for the Defense Environmental Restoration Program* (DERP), dated September 2001 (provided to GAO on July 21, 2008 and February 12, 2009)
- Final Rule of the Munitions Response Site Prioritization Protocol (32 CFR Part 179), dated October 2005 (GAO extracted the document from the DENIX web site)

Currently, the DoD Components must notify OSD of significant environmental events involving compliance with environmental statutes, environmental enforcement actions, and chemical emergencies or spills. Furthermore, the DoD Components must notify the public if a public water system does not meet the Safe Drinking Water Act standards. If the DoD Components obtain new information about a previously known release, they are already required to review and evaluate

See comment 4.
See comment 5.

98 United States Government Accountability Office

any potential impacts to the cleanup process in consultation with relevant stakeholders, to include regulators.

RECOMMENDATION 4: The GAO recommends that the Secretary of Defense ensure that the Services make a determination of appropriateness using Office of Management and Budget criteria and Service guidance, before using performance-based contracts (PBCs) for Superfund cleanup. (See page 39/GAO Draft Report.)

DoD RESPONSE: DoD concurs. In June 2007, OSD released the internal publication entitled *Performance-Based Acquisition of Environmental Restoration Services*, which addresses the suitability of PBCs. The handbook specifically states that PBCs "may not be appropriate for all environmental restoration projects," especially site characterization. It identifies risk and uncertainty as the primary considerations in determining the suitability of a project for PBC and elaborates on these issues.

RECOMMENDATION 5: The GAO recommends that the Secretary of Defense direct the Services to document compliance with relevant EPA guidance when selecting monitored natural attenuation to ensure that DoD NPL sites utilize monitored natural attenuation as the sole remedy at contaminated sites only when it is documented to meet remediation objectives that are protective of human health and the environment. (See page 39/GAO Draft Report.)

DoD RESPONSE: DoD concurs. DoD provided guidance on remedy selection and monitored natural attenuation in the *Management Guidance for the DERP*, dated September 2001 (provided to GAO on July 21, 2008 and February 12, 2009). The DERP guidance requires the DoD Components to consider appropriate treatment technologies, permanent solutions, containment strategies, land use controls, and alternate water supplies when evaluating groundwater remedial alternatives during the Feasibility Study phase. Monitored natural attenuation (MNA) may be selected as the preferred remedial alternative only if the site conditions support MNA as a viable remedy. The DoD Components document the preferred remedial alternative in a proposed plan, along with a brief description of the remedial alternatives evaluated. Regulators and the public review and comment on the proposed plan. Once a remedial action is selected, the DoD Component prepares a Record of Decision (ROD). The ROD defines the remedial action objectives and describes how the selected remedy is protective of human health and the environment. At NPL sites, EPA must sign the ROD, and thus concur that the remedy is protective of human health and the environment.

RECOMMENDATION 6: The GAO recommends that the Administrator of EPA establish a record-keeping system for DOD NPL sites, consistent across all

See comment 6.

Superfund: Interagency Agreements and Improved Project...   99

regions, to accurately track documents submitted for review, including the status of approvals.

DoD RESPONSE: DoD concurs. DoD agrees that accurate data and effective documentation management is critical to EPA's ability to provide oversight of cleanup at NPL installations. As part of the joint effort to improve communication and better correlate reporting of cleanup progress (see DoD's response to Recommendation 1), DoD and EPA are currently reviewing information in and the capabilities of existing databases.

Matter for Congressional Consideration: The GAO recommends that Congress may want to consider amending section 120 of CERCLA to authorize EPA—after an appropriate notification period—to administratively impose penalties to enforce cleanup requirements at federal facilities.

DoD RESPONSE: DoD nonconcurs. DoD does not agree that Congress should consider amending section 120 of CERCLA to authorize EPA to administratively impose penalties to enforce cleanup requirements at federal facilities without a negotiated CERCLA IAG for the following reasons:

- EPA currently has existing statutory enforcement tools, such as imminent and substantial endangerment orders under RCRA or the SDWA.
- EPA has authority to negotiate administrative penalties in IAGs under CERCLA. DoD-EPA IAGs have included stipulated penalties for a number of years. Providing EPA the authority to issue CERCLA penalties at facilities without an IAG may prove to be a disincentive to EPA *negotiating* these interagency agreements.
- EPA has significant authority at NPL installations without administrative orders and penalties. EPA has remedy selection authority at NPL installations regardless of whether the installation has a signed IAG.

We are pleased to note that 136 of 141 DoD-EPA FFAs have been finalized and only 5 remain in negotiation. DoD is committed to continuing its progress on negotiating FFAs with EPA under the current framework.

The following are GAO's comments on the Department of Defense's letter, dated July 5, 2010.

GAO Comments

1. For this recommendation DOD agreed that it is vitally important to track cleanup progress at its installations and to make that information available to Congress and the public in a manner that is transparent

and easily understandable. DOD also discussed working actively with EPA through a federal working group. However, DOD indicated that if the working group decides a common metric is essential, DOD would require that the metric meet DOD criteria, such as continuing use of DOD's site level measure as compared to EPA's operating unit level of measure, suggesting the agencies are unlikely to implement a uniform method for reporting cleanup progress at the installations. We continue to believe that such uniformity is essential to provide greater assurance that cleanup progress is being measured accurately and consistently across all Superfund sites, and to provide for transparency to Congress and the public. An agency may need more detailed information for management purposes, but information comparable to other Superfund sites is essential to providing adequate transparency.

2. DOD disagreed with our recommendation that EPA pursue amendments to Executive Order 12580 to condition delegation of CERCLA authorities to DOD on the existence of a signed IAG. DOD stated that because all but 5 of the 141 IAGs remain in negotiation, DOD should maintain lead agency CERCLA authority so it can continue executing cleanup actions pending resolution of any IAG issues and indicated its intention to sign the remaining 5 IAGs using as a template an IAG between the Army and EPA for Fort Eustis, Virginia, as has been agreed upon by the agencies. However, given that 4 remaining agreements have been pending for over a decade, we continue to believe that outstanding CERCLA Section 120 IAGs need to be negotiated expeditiously and that amendments to Executive Order 12580 could facilitate such action.

3. For this recommendation, the Deputy Under Secretary agreed that proper notification of new releases that exceed statutory thresholds and significant new information about previously known releases is necessary. DOD noted that DOD guidance on this issue is already in existence; however, GAO reviewed these documents during the engagement and found them to lack adequate specificity for use by installation personnel, particularly in the area of new information about previous releases. Although the Deputy Under Secretary notes that when DOD personnel obtain new information about a previously known release they are already required to review and evaluate any potential impacts to the cleanup process in consultation with relevant stakeholders, to include regulators, we found several instances where

Superfund: Interagency Agreements and Improved Project...   101

DOD personnel did not share such information with regulators in a timely fashion. When we asked why, installation personnel stated they were not required to provide regulators with such information. For example, our report highlights an example of Tyndall's failure to notify EPA about the presence of lead—a hazardous substance under CERCLA—at the Tyndall elementary school, and failure to take action to prevent children's exposure to lead shot, among other issues.

4. For this recommendation, DOD agreed and referenced its policy. However, our review found inconsistencies in how this policy was interpreted. While federal guidelines indicate that performance-based contracts (PBC) are not generally appropriate for work that involves a great deal of uncertainty, officials from the Army told us that in their view, PBCs are better suited for complex work because they foster innovation from the private sector. DOD policy directs the services to use PBCs whenever possible—establishing the goal that PBCs be used for 50 percent of service acquisitions. Nonetheless, Tyndall AFB officials told us that after shifting toward PBCs for cleanup work in 2004, they are no longer using them for new contracts because of the uncertainty in the cleanup work needed at the base.

5. For this recommendation, DOD agreed and referenced its DERP guidance, which outlines the process for developing and proposing remedies. The guidance, however, does not provide specific requirements regarding monitored natural attenuation. As DOD notes, when DOD selects monitored natural attenuation as its remedy, DOD is to present the basis for its selection in a ROD or proposed plan. However, DOD and its contractors are not uniformly demonstrating that EPA's specific criteria for selection of monitored natural attenuation are met before selecting such a remedial alternative, according to EPA. These criteria require that certain conditions exist such as a low potential for contaminant migration and a time frame comparable to other methods of remediation.

6. The Deputy Under Secretary of Defense disagreed with our Matter for Congressional Consideration, in which we suggested that Congress should consider amending section 120 of CERCLA to authorize EPA to administratively impose penalties to enforce cleanup requirements at federal facilities without a negotiated CERCLA interagency agreement. DOD presented several reasons for its position, including its belief that EPA has existing statutory enforcement tools under the Resources Conservation Recovery Act (RCRA) and the Safe Drinking

Water Act (SDWA). However, there is little evidence that these other mechanisms have been effective. For example, in 2007 EPA issued administrative cleanup orders under RCRA at all three installations that the services disagreed with and they all initially refused to comply while DOD sought DOJ review of the orders' validity. The orders stated that an imminent and substantial endangerment from contamination may be present on the sites and required DOD to notify EPA of its intent to comply and clean up. The Air Force and Army did not notify EPA of their intent to comply with the order within the time frame required and stated they would continue to clean up the sites under their CERCLA removal and lead agency authority. After DOJ issued a letter stating its opinion that EPA had the authority to issue the orders, as a matter of law, the Army informed EPA of its intent to comply and initiated work under RCRA at Fort Meade, while the Air Force did not take similar actions for its installations. Subsequent negotiations between DOD and EPA resulted in IAGs at Fort Meade and McGuire AFB. However, at Tyndall AFB, where there is still no signed IAG, DOD continues to refuse to comply with the RCRA order. In regards to SDWA, we recognize there can be installations with contamination that do not threaten a public water supply, and therefore SDWA would not apply. DOD also commented that EPA has authority to negotiate administrative penalties in IAGs under CERCLA and that existing IAGS include stipulated penalties. However, as we stated previously, several of the most challenging sites do not yet have IAGs, including Tyndall AFB. For more than a decade DOD has failed to enter into IAGs required by CERCLA section 120 to clean up DOD National Priorities List (NPL) sites. As we note in our report, without an IAG EPA lacks the mechanisms to ensure that cleanup by an installation proceeds expeditiously, is properly done, and has public input, as required by CERCLA. We disagree that providing EPA with the authority to issue CERCLA penalties at facilities without an IAG will be a disincentive to EPA's negotiating interagency agreements. EPA has stated on numerous occasions its commitment to complete negotiations for such agreements. Finally, DOD noted that EPA has remedy selection authority at NPL installations regardless of whether the installation has a signed IAG. Despite having authority for choosing a final cleanup remedy, EPA has not been able to force progress toward remedy selection because it has no enforceable schedule to ensure

DOD installations make progress on the technical steps leading up to the ROD, which documents the remedy selected for cleanup. Hence, as at the three installations reviewed in this chapter, installations may not complete cleanup for a decade or more without an IAG. We believe our report demonstrates that EPA has experienced considerable difficulty employing its existing enforcement authorities and that DOD has resisted EPA's use of such authority to compel DOD to enter into IAGs at NPL sites. Hence, we continue to assert that an expansion in EPA's enforcement authority is warranted.

GLOSSARY

This glossary is provided for reader convenience. It is not intended as a definitive, comprehensive glossary of all aspects of the Comprehensive Environmental Response, Compensation, and Liability Act (CERCLA) process for the cleanup of environmental contamination at Superfund sites.

Site Discovery

When a federal agency identifies an actual or suspected release or threatened release to the environment on a federal site, it notifies EPA, which then lists the site on its Federal Agency Hazardous Waste Compliance Docket. The docket is a listing of all federal facilities that have reported hazardous waste activities under RCRA or CERCLA. RCRA and CERCLA require federal agencies to submit to EPA information on their facilities that generate, transport, store, or dispose of hazardous waste or that has had some type of hazardous substance release or spill. EPA updates the docket periodically.

Preliminary Assessment

The lead agency (DOD, in this case) conducts a preliminary assessment of the site by reviewing existing information, such as facility records, to determine whether hazardous substance contamination is present and poses a potential threat to public health or the environment. EPA regions review these preliminary assessments to determine whether the information is sufficient to

104   United States Government Accountability Office

the likelihood of a hazardous substance release, a contamination pathway, and potential receptors. EPA regions are encouraged to complete their review of preliminary assessments of federal facility sites listed in EPA's CERCLA database within 18 months of the date the site was listed on the federal docket. EPA may determine the site does not pose a significant threat to human health or the environment and no further action is required. If the preliminary assessment indicates that a long-term response may be needed, EPA may request that DOD perform a site inspection to gather more detailed information.

Site Inspection

The lead agency (DOD, in this case) samples soil, groundwater, surface water, and sediment, as appropriate, and analyzes the results to prepare a report that describes the contaminants at the site, past waste handling practices, migration pathways for contaminants, and receptors at or near the site. EPA reviews the site inspection report and, if it determines the release poses no significant threat, EPA may eliminate it from further consideration. If EPA determines that hazardous substances, pollutants, or contaminants have been released at the site, EPA will use the information collected during the preliminary assessment and site inspection to calculate a preliminary HRS score.

Hazard Ranking System Scoring

If EPA determines that a significant hazardous substance release has occurred, the EPA region prepares an HRS scoring package. EPA's HRS assesses the potential of a release to threaten human health or the environment by assigning a value to factors such as (1) the likelihood that a hazardous release has occurred; (2) the characteristics of the waste, such as toxicity and the amount; and (3) people or sensitive environments affected by the release.

National Priorities List

If the release scores an HRS score of 28.50 or higher, EPA determines whether to propose the site for placement on the NPL. CERCLA requires EPA to update the NPL at least once a year.

Remedial Investigation and Feasibility Study

Within 6 months after EPA places a site on the NPL, the lead agency (DOD, in this case) is required to begin a remedial investigation and feasibility study to assess the nature and extent of the contamination. The remedial investigation and feasibility study process includes the collection of data on site conditions, waste characteristics, and risks to human health and the environment; the development of remedial alternatives; and testing and analysis of alternative cleanup methods to evaluate their potential effectiveness and relative cost. EPA, and frequently the state, provides oversight during the remedial investigation and feasibility study and the development of a proposed plan, which outlines a preferred cleanup alternative. After a public comment period on the proposed plan, EPA and the federal facility sign a record of decision (ROD) that documents the selected remedial action cleanup objectives, the technologies to be used during cleanup, and the analysis supporting the remedy selection.

Interagency Agreement

Within 6 months of EPA's review of DOD's remedial investigation and feasibility study, CERCLA, as amended, requires that DOD enter into an IAG with EPA for the expeditious completion of all remedial action at the facility. (EPA's policy however, is for federal facilities to enter into an IAG after EPA places the site on the NPL.) The IAG is an enforceable document that must contain, at a minimum, three provisions: (1) a review of remedial alternatives and the selection of the remedy by DOD and EPA, or remedy selection by EPA if agreement is not reached; (2) schedules for completion of each remedy; and (3) arrangements for the long-term operation and maintenance of the facility.

Remedial Design and Remedial Action

During the remedial design and remedial action process, the lead agency (DOD, in this case) develops and implements a permanent remedy on the site as outlined in the record of decision and IAG.

Monitoring

Long-term monitoring occurs at every site following construction of the remedial action. This includes the collection and analysis of data related to chemical, physical, and biological characteristics at the site to determine whether the selected remedy meets CERCLA objectives to protect human health and the environment. For NPL or non-NPL sites where hazardous substances, pollutants, or contaminants were left in place above levels that do not allow for unlimited use and unrestricted exposure, every 5 years following the initiation of the remedy, the lead agency (DOD, in this case) must review its sites. The purpose of a 5-year review, similar to long-term monitoring, is to assure that the remedy continues to meet the requirements contained in the record of decision and is protective of human health and the environment.

End Notes

[1] The environmental restoration expenditures total does not include program management and other support costs. Under its Defense Environmental Restoration Program, DOD's authority for environmental cleanup includes each facility or site owned by, leased to, or otherwise possessed by the United States and under the jurisdiction of DOD, as well as those that were as such at the time of actions leading to contamination by hazardous substances or other hazards prior to October 17, 1986. DOD notes that this jurisdiction extends to governmental entities that are the legal predecessors of DOD or the components—Army, Navy, Marine Corps, and Air Force.

[2] Pub. L. No. 96-510 (1980), codified as amended at 42 U.S.C.§§ 9601-9675 (2010).

[3] The NPL is composed of 1,279 final sites and 341 deleted sites. There are an additional 61 proposed sites.

[4] GAO, *Superfund: Greater EPA Enforcement and Reporting Are Needed to Enhance Cleanup at DOD Sites*, GAO-09-278 (Washington, D.C.: Mar. 13, 2009).

[5] CERCLA, Pub. L. No. 96-510 § 120(e), as amended by the Superfund Amendments and Reauthorization Act, Pub. L. No. 99-499 (1986); codified at 42 U.S.C. § 9620(e) (2010).

[6] For purposes of this chapter, the term "installation" refers generally to a property under the jurisdiction of DOD, and for which it has cleanup responsibility, or specifically to one of the three properties under review. DOD has other definitions for the term.

[7] We refer to Fort George G. Meade as "Fort Meade" throughout this chapter.

Superfund: Interagency Agreements and Improved Project... 107

[8] Each of the four installations received administrative orders where an imminent and substantial endangerment to health and the environment may exist under the authority of the Resource Conservation and Recovery Act or under the authority of the Safe Drinking Water Act.

[9] Although Air Force Plant 44 is near completion of the cleanup required under the Safe Drinking Water Act, the IAG remains unsigned.

[10] Pub. L. No. 94-580 (1976), amending the Solid Waste Disposal Act, codified as amended at 42 U.S.C. §§ 6921 – 6992k (2010).

[11] Section 2 of RCRA added section 7003 to the Solid Waste Disposal Act, but the imminent hazard authority is nonetheless often referred to as "RCRA Section 7003." See 42 U.S.C. § 6973 (2010).

[12] Applicable or relevant and appropriate requirements include standards promulgated under any federal environmental law, in addition to standards promulgated under certain state laws or regulations that are more stringent than corresponding federal law and are identified to the entity leading the cleanup in a timely manner. See National Oil and Hazardous Substances Pollution Contingency Plan, 40 C.F.R. Pt. 300 (2010).

[13] For purposes of this chapter, we refer to "removals" as defined in the NCP and EPA's Superfund Program Implementation Manual, and as distinct from the other steps such as the remedial investigation and feasibility study (RI/FS). However, EPA notes that in some contexts, the agency considers removals to include all phases of work from preliminary assessment through the RI/FS.

[14] The National Response Center is the sole federal point of contact for reporting all hazardous substances and oil spills that trigger federal notification requirements under several laws. Information reported to the Center is disseminated to other agencies, such as EPA, as well as to states.

[15] 42 U.S.C. §§ 9620(e)(4) (2010).

[16] As we explain in this chapter, by agreement of the agencies, all IAGs between EPA and DOD entered after February 2009 follow as a template the IAG executed in March 2008 by the Army and EPA for Fort Eustis, Va. Our general description of IAGs is based on the Fort Eustis template, although some features were also included in IAGs predating this template. For example, site management plans have been included in IAGs since 1999.

[17] The key difference is that for these and other secondary documents there is no provision for dispute resolution, and thus DOD need not address all EPA comments to EPA's satisfaction. For purposes of this chapter, we refer to the approvals subject to dispute resolution as formal EPA approval.

[18] Any removal action should, to the extent practicable, contribute to the efficient performance of any long-term remedial action with respect to the release or threatened release concerned. 42 U.S.C. § 9604(a)(2) (2010). For example, the IAG for McGuire AFB establishes that the agreement does not affect the Air Force authority under CERCLA section 104 to conduct removals. Under the IAG, removal-related documents such as Non-Time Critical Removal Action Plans and Removal Action Memoranda, are secondary documents. While the IAG provides that the Air Force and EPA have certain obligations regarding submission, review, and response to comments for such documents, these documents are not expressly subject to dispute resolution unless they are feeder or input documents to a primary document, such as a remedial action. According to DOD officials, this distinction means that the federal facility may conduct a removal action without formal concurrence from EPA. Nonetheless, EPA officials stated that due to its authorities for remedy selection, it has an interest in ensuring, at a minimum, that a removal action does not interfere with a remedial action work plan, as included in the Fort Eustis IAG template.

[19] H.R. Rep. No. 99-253, pt. 1 at 95 (1985).

[20] H.R. Conf. Rep. No. 99-962 at 242 (1986).

[21] CERCLA's citizen suit provision, codified at 42 U.S.C. § 9659, authorizes such suits to enforce any standard, regulation, condition, requirement, or order, including any provision of an IAG, effective under CERCLA. At a federal NPL site without an IAG, this provision would

108 United States Government Accountability Office

grant a right to sue where a federal agency has violated, for example, a ROD, to the extent it contains an enforceable standard, condition, or requirement; however, judicial review regarding choice of a remedy could not occur until after all activities in the ROD were completed. Moreover, as discussed later in this chapter, sites without IAGs may not achieve RODs, and/or may take many years to achieve RODs, thus limiting the role of citizen suits as a pragmatic enforcement mechanism for such sites. At such sites, a federal agency could either delay cleanup or IAGs indefinitely, without risk of a citizen suit or conduct removals without RODs.

[22] Formerly used defense sites are located on properties that were under the jurisdiction of DOD and owned by, leased to, or otherwise possessed by the United States prior to October 17, 1986, but have since been transferred to states, local governments, other federal entities, or private parties. See 10 U.S.C. § 2701(c)(1)(B) (2010); see also footnote 1.

[23] 10 U.S.C. § 2705(a) (2010) ("Expedited notice").

[24] E.O. No. 12580, 52 Fed. Reg. 2923 (Jan. 23, 1987).

[25] The E.O. reflects this, stating that "[t]he conduct and control of all litigation arising under [CERCLA] shall be the responsibility of the Attorney General." Exec. Order No. 12,580 § 6(a).

[26] See Environmental Compliance by Federal Agencies: Hearing Before the Subcommittee on Oversight and Investigations of the House Committee on Energy and Commerce, 100th Congress 668, 675 (1987) (memorandum from John Harmon, Assistant Attorney General, Office of Legal Counsel, to Michael J. Egan, Associate Attorney General, June 23, 1978) (stating DOJ view that allowing EPA to sue another agency would violate established principle that "no man can create a justiciable controversy against himself").

[27] Pub. L. No. 102-386.

[28] GAO-09-278.

[29] DOD notes that when it proceeded with removals for which it has authority to proceed without formal EPA approval, it did so in order to protect human health and the environment. These types of actions typically are used to reduce immediate risks, and do not replace the full CERCLA process which ensures long-term protectiveness and is subject to formal EPA approval via RODs and other documents.

[30] RCRA provides EPA with the authority to issue administrative enforcement orders to address solid and hazardous wastes that may present an imminent and substantial endangerment to public health or the environment.

[31] The Safe Drinking Water Act provides EPA with authority to take action to protect human health from contamination present or likely to enter a public water system which may present an imminent and substantial endangerment.

[32] DOJ stated that EPA may require DOD to agree in the IAG to follow, "in the same manner and to the same extent" as they apply to private parties, any "guidelines, rules, regulations, and criteria" established by EPA and made applicable to nonfederal facilities under CERCLA, noting that EPA's model agreements for federal facilities and for private parties may inform the content of such terms.

[33] Letter, William C. Anderson, Assistant Secretary for Installations, Environment and Logistics, Department of the Air Force, to Granta Nakayama, Assistant Administrator for Office of Enforcement and Compliance Assurance, EPA, May 28, 2008.

[34] While the cleanup processes under CERCLA and RCRA have many analogous steps, there are distinctions between the two regulatory processes. DOD notes that in other RCRA contexts, such as closure of a RCRA-permitted waste facility, EPA has recognized cleanup work conducted under CERCLA, consistent with EPA guidance.

[35] CERCLIS is the database and data management system that EPA uses to track activities at Superfund sites.

[36] U.S. Department of Defense. *Fiscal Year 2008 Defense Environmental Programs Annual Report to Congress.* (Washington, D.C., July 16, 2009).

Superfund: Interagency Agreements and Improved Project... 109

[37] DOD's term "response complete" means the remedy is in place and required remedial action operations, if any, have been completed. DOD categorizes as "response complete" sites where the agency has determined no cleanup remedy is required, as well as sites where a cleanup remedy has been fully implemented. See also DOD DERP Guidance (2001).

[38] GAO, *Military Munitions Response Program: Opportunities Exist to Improve Program Management*, GAO-10-384 (Washington, D.C.: Apr. 9, 2010); *Environmental Contamination: Cleanup Actions at Formerly Used Defense Sites*, GAO-01-557 (Washington, D.C.: July 31, 2001).

[39] In addition, EPA provided us with a copy of other documents developed by the installations, such as cleanup schedules, which also included planned actions anticipated in the near future.

[40] The DERP statute directs DOD to carry out its hazardous substances cleanup program in accordance with CERCLA, and CERCLA is DOD's preferred framework for environmental restoration.

[41] DOD's Annual Report to Congress for Fiscal Year 2008 defines the "study" phase as comprised of three investigation phases: preliminary assessment, site inspection, and RI/FS.

[42] According to EPA officials, DOD developed its own environmental reporting metrics without consulting EPA.

[43] In commenting on this chapter, DOD stated that it believes it is difficult to obtain EPA concurrence on cleanup decisions at such sites because of lack of resources, delays in review, and sometimes disagreement over issues specific to the ROD.

[44] See 42 USC § 9620(e)(4)(B)-(C) (2010) (providing that IAGs shall contain "[a] schedule for the completion of each such remedial action [and a]rrangements for long-term operation and maintenance of the facility").

[45] DOD responded to EPA comments in a letter approximately 1 year later.

[46] In commenting on this chapter, DOD acknowledged that the volume of submissions may vary substantially over time, but stated that schedules prepared for other purposes—such as DOD agreements with states—typically identify time frames for submittal of many of these documents.

[47] CERCLA requires specific community involvement activities that must occur at certain points throughout the cleanup process. These activities include, but are not limited to, public meetings, requests for public comment, and availability of Superfund decision documents. 42 U.S.C. §§ 9613(k), 9620(e)(2), 9617(a)-(b), 9620(f), 9621(f)(1) (2010).

[48] 48 C.F.R. 37.102(a)(1)(i) (FAR) (2010).

[49] DOD Guidebook (2000), available at https://www.acquisition.gov/SevenSteps/library/DODguidebook-pbsa.pdf. (last accessed May 26, 2010).

[50] Air Force Center for Environmental Excellence, Environmental Restoration Performance Based Contracting (PBC) Concept of Operations (February 2007), p. 12-14.

[51] EPA issued a memorandum in 2006 regarding EPA's role and responsibilities with respect to federal agencies' use of PBCs for federal facility cleanups. See OSWER Guidance 9272.0-21. The memorandum reflects the federal government preference for PBCs, while observing that federal agencies using PBCs may tend to provide less oversight of contractors than they had using traditional contracts, among other concerns. The memorandum also states that PBCs should clearly define performance objectives, and that general objectives such as "work with regulators to obtain approval" are not appropriate.

[52] VOCs are emitted as gases from certain solids or liquids. VOCs include substances—some of which may have short- and long-term adverse health effects—such as benzene, toluene, methylene chloride, and methyl chloroform.

[53] PCBs belong to a broad family of synthetic organic chemicals known as chlorinated hydrocarbons. PCBs were domestically manufactured from 1929 until their manufacture was banned in 1979. PCBs have been demonstrated to cause a variety of adverse health effects, including cancer and other serious non-cancer effects.

110 United States Government Accountability Office

[54] The PA/SI is used by EPA as well as the lead agency to evaluate whether the site may pose a threat to public health or the environment and whether there is any potential need for removal action, and to collect data to evaluate the release of hazardous substances from a site.

[55] AR 200-1 § 15-5 (2007).

[56] 10 U.S.C. § 2705(a) (2010) ("Expedited notice").

[57] Paragraph 66(b) of the order states that "[i]n the event of any release of a hazardous substance from the facility, [Tyndall AFB] shall immediately notify" the EPA RPM, EPA Region 4, and the National Response Center, and submit a written report within 7 days.

[58] The U.S. Department of Health and Human Services ATSDR is a federal public health agency that is charged by CERCLA to assess the presence of health hazards at Superfund sites and to provide information about risks relevant to the need to reduce further exposure to those hazards. This requirement is not limited to the threshold reportable quantities established in CERCLA regulations.

[59] For copies of ATSDR's reports for Tyndall, see:
http://www.atsdr.cdc.gov/HAC/PHA/HCPHA.asp?State=FL

[60] Munitions constituents are defined as any materials originating from unexploded ordnance, discarded military munitions, or other military munitions, including explosive and nonexplosive materials, and emission, degradation, or breakdown elements of such ordnance or munitions (10 U.S.C. § 2710 (e)(4) (2010).

[61] The Phase I report identified the Stationary Target Range as one such area, and noted that the Tyndall Elementary School is located on a portion of the former range. The report further stated that lead shot was observed on the ground in some places, and that lead shot had previously been found at the school, but did not state specifically whether the school property was inspected.

[62] The CSE Phase 1 Report was provided to EPA in mid-2007, but Tyndall proceeded before receiving EPA's review, which was provided later in 2007, or concurrence.

[63] According to EPA, Tyndall's contractor told EPA they observed lead shot and clay target debris on the playground during a kickoff site walk in August 2008, but the contractor denied this when questioned by GAO. While the officials did not enter the school property during the site walk, they were adjacent to the property and could see the ground through or over a chain link fence at the property boundary. Further, by AFCEE's March 2009 visit, Tyndall representatives acknowledged they had recently observed lead shot at the school.

[64] According to Air Force Center for Engineering and the Environment officials, Tyndall AFB is responsible for any hazardous substances response at the school, even though it is leased to the county.

[65] In contrast, Fort Meade officials conducted periodic surveys to ensure that the Army identified any unexploded ordnance at the surface of the Patuxent National Wildlife Refuge.

[66] In commenting on this chapter, DOD disagreed with GAO's characterization of the ATSDR report, and asserts that the report found that the lead shot did not present a health hazard. However, ATSDR officials based the health assessment on their understanding that no ongoing exposure to the shot itself was occurring.

[67] DOD notes that according to ATSDR guidance, EPA should have received a copy of the 2000 ATSDR report, and asserts that this should have alerted EPA to the presence of lead pellets at the school.

[68] EPA officials stated that a preliminary assessment, and potentially a removal action, was indicated based on the data Tyndall AFB had in 1992. Similarly, once Tyndall officials observed the lead shot on the playground at some time prior to March 2009, Tyndall should have undertaken further investigation right away, which would have led to soil sampling and the removal as were eventually conducted, as well as short-term measures to prevent children's exposure. ATSDR officials told us that if they knew that lead shot remained in the playground after 1992, they would have assessed its risk in their 2000 assessment.

[69] E.O. No. 12580 § 4(e) (1987).

Superfund: Interagency Agreements and Improved Project... 111

[70] In commenting on this chapter, DOD noted that the requirement for DOJ concurrence provides a check on EPA authority which DOD asserts is analogous to a private party's right to challenge an EPA order in court—a mechanism not available to federal agencies for the same reasons that EPA cannot bring a federal agency into court to enforce an order. DOD further notes that EPA has "informal tools" such as political pressure and interagency dispute mechanisms. However, GAO believes that in some instances—such as the three sites studied here, where DOD has failed to enter IAGs for over a decade—these tools are insufficient given EPA's special role as the regulator—rather than a response agency— under CERCLA.

[71] Where there has been a release of a hazardous substance where DOD is the lead agency, CERCLA section 103(a) requires DOD to report such releases above reportable quantities to the National Response Center. CERCLA section 111(g), as delegated by E.O. 12580 § 8(b), requires DOD to promulgate rules and regulations regarding notification of potentially injured parties of such release, and until such promulgation, requires reasonable notice to potential injured parties by publication in local newspapers serving the affected area. Finally, 10 U.S.C. § 2705(a) requires the Secretary of Defense to take necessary actions to ensure that EPA and state authorities receive prompt notice of the discovery of a release or threatened release, the associated extent of the threat to public health and the environment, proposals to respond to such release, and initiation of any response.

[72] GAO-09-278.

[73] As previously mentioned these four sites include the Defense Reutilization and Marketing Office, Closed Sanitary Landfill, Clean Fill Dump, and Post Laundry Facility.

[74] The signatories of the Federal Facility Agreement/IAG for Fort Meade include EPA Region 3, the Army, Department of the Interior, and Architect of the Capitol.

[75] The number of sites is as of the end of 2008.

INDEX

A

abatement, 70
accountability, 34, 78
acquisitions, 67, 101
adhesives, 43
Afghanistan, 77
agencies, 4, 5, 7, 10, 11, 12, 13, 14, 15, 18,
    19, 21, 22, 26, 27, 34, 35, 46, 49, 51, 52,
    53, 54, 55, 56, 58, 63, 69, 72, 76, 78, 79,
    82, 100, 103, 107, 109, 111
Air Force, 21, 22, 24, 25, 26, 29, 30, 31, 43,
    46, 47, 48, 50, 56, 57, 59, 65, 66, 68, 71,
    72, 73, 74, 75, 77, 80, 82, 83, 84, 87, 88,
    89, 92, 102, 106, 107, 108, 109, 110
Alaska, 12
appropriations, 10, 42, 51
aquifers, 86, 90
aromatic hydrocarbons, 83
arsenic, 84, 85, 90
assessment, vii, 1, 5, 6, 7, 11, 13, 15, 16, 19,
    25, 27, 28, 33, 34, 35, 37, 41, 47, 53, 54,
    64, 70, 73, 75, 103, 104, 107, 109, 110
Attorney General, 54, 108
authorities, viii, 2, 5, 7, 12, 13, 26, 27, 37,
    41, 42, 45, 49, 51, 54, 57, 79, 80, 100,
    103, 107, 111

B

background, vii, 43
barium, 85, 90
benzene, 90, 109

Boeing, 88

C

cadmium, 83
cancer, 44, 109
challenges, 34, 56, 77, 90
chlorinated hydrocarbons, 109
chloroform, 84, 109
chromium, 83, 84, 90
City, 89
civil action, 55
clarity, 61, 78
cleaning, viii, 2, 3, 6, 9, 21, 30, 31, 45, 49,
    78
Coast Guard, 13
communication, 63, 78
community, 12, 109
community support, 12
compliance, 12, 21, 22, 25, 42, 43, 47, 49,
    54, 57, 63, 76, 77, 78, 79, 80, 87, 89, 91
compounds, 83, 86, 88
conference, 11, 42, 53
conflict, 8, 22, 67
consent, 43
consulting, 109
contaminant, 66, 101
contamination, vii, 4, 5, 6, 8, 13, 14, 15, 16,
    22, 23, 25, 27, 29, 30, 31, 33, 37, 38, 44,
    47, 56, 57, 58, 62, 65, 66, 68, 69, 71, 72,
    73, 76, 78, 86, 88, 90, 91, 102, 103, 105,
    106, 108
coordination, 65, 69
copper, 85

114 Index

copyright, iv
Copyright, iv
cost, 16, 17, 52, 68, 69, 105
counsel, 57
crude oil, 42
cyanide, 30

D

database, 7, 13, 15, 35, 43, 60, 61, 62, 104, 108
degradation, 110
delegates, 41, 42, 54
Department of Agriculture, 13
Department of Defense, 1, iii, vii, viii, 1, 3, 4, 35, 40, 41, 45, 47, 49, 80, 82, 87, 90, 94, 99, 108
Department of Energy, 13, 43
Department of Health and Human Services, 43, 110
Department of Justice, 3, 8, 47, 51
Department of the Interior, 13, 82, 111
destination, 19, 43
draft, 9, 22, 69, 70, 88, 91
drinking water, 31, 43, 90
dyes, 44

E

economic development, 29, 30
Education, 90
electroplating, 30
elementary school, 70, 101
eligibility criteria, 44
emergency response, 51
enforcement, 2, 4, 6, 8, 9, 12, 18, 19, 21, 22, 23, 24, 26, 33, 35, 42, 43, 54, 55, 76, 101, 108
engineering, 23, 67
environmental conditions, 70, 75
environmental contamination, 103
Environmental Protection Agency, vii, viii, 1, 3, 4, 35, 39, 45, 47, 49, 80, 82, 87, 90, 92
equipment, vii, 4, 31, 44, 49
exclusion, 35

Executive Order, 41, 42, 43, 51, 54, 100
exercise, 26, 89
exposure, 10, 18, 33, 51, 73, 75, 101, 106, 110

F

Federal Aviation Administration, 13
federal law, 71, 76, 107
fish, 72, 90
Fish and Wildlife Service, 50, 72, 81, 90
flexibility, 68, 69
flight, 30, 90
fluid, 84
funding, 4, 23, 44, 67, 74

G

geography, 19, 90
greed, 29
groundwater, 15, 19, 29, 30, 31, 47, 56, 58, 66, 68, 70, 86, 90, 104
guidance, 19, 35, 36, 68, 79, 81, 100, 101, 108, 110
guidelines, 26, 42, 50, 67, 68, 81, 101, 108

H

Hawaii, 24, 43, 59, 82, 84
hazardous materials, 14, 28
hazardous substances, viii, 2, 5, 7, 10, 16, 18, 19, 20, 30, 41, 43, 45, 49, 50, 51, 56, 58, 64, 72, 80, 81, 84, 86, 104, 106, 107, 109, 110
hazardous wastes, vii, 4, 86, 88, 91, 108
hazards, 26, 38, 70, 106, 110
headquarters, 6, 8, 20, 22, 25, 27, 28, 29, 35, 36, 67
health effects, 75, 109
heavy metals, 70, 86
host, 10, 51, 87
housing, 65, 69
hydraulic fluids, 84
hydrocarbons, 88

Index

I

impacts, 100
inclusion, 3, 13, 16, 28, 35, 36, 87
initiation, 18, 54, 106, 111
inspections, 38, 42, 54
intervention, 66
Iraq, 77
issues, vii, 22, 35, 65, 67, 78, 86, 100, 101, 109

J

jurisdiction, 33, 42, 87, 106, 108
justification, 34

L

landfills, 30, 31, 90
leaching, 30, 75
leadership, 55
legislation, vii, viii, 1, 4, 35, 45, 49
litigation, 18, 22, 55, 76, 108
local government, 108
lubricating oil, 84

M

majority, 2, 7, 10, 26, 27, 28, 38
management, 3, 9, 10, 21, 23, 28, 35, 43, 48, 52, 63, 64, 65, 67, 79, 81, 88, 91, 100, 106, 107, 108
manufacturing, 49
Marine Corps, 106
memorandums of understanding, 26
mercury, 83, 84, 85, 88
metabolites, 90
metals, 30, 88
methodology, 7, 36, 38, 50
methylene chloride, 109
migration, 15, 31, 66, 101, 104
military, vii, 4, 7, 23, 24, 31, 36, 37, 49, 69, 77, 82, 87, 91, 110

N

National Aeronautics and Space Administration, 13
National Defense Authorization Act, 37
natural resources, 49, 77
negotiating, 24, 102
nervous system, 43, 44
nickel, 84, 88

O

obstacles, viii, 45, 46, 50, 65, 80, 81
Office of Management and Budget, 3, 9, 79
oil, 31, 90, 107
oil spill, 107
omission, 70
organic chemicals, 109
organic compounds, 70, 86, 88
overlap, 60
oversight, vii, viii, 1, 2, 6, 7, 9, 10, 12, 17, 19, 22, 26, 27, 34, 35, 37, 42, 43, 45, 49, 52, 56, 72, 73, 80, 105, 109

P

Pacific, 84
Panama, 89
pathways, 15, 104
penalties, 11, 18, 19, 21, 22, 23, 34, 54, 55, 57, 67, 76, 80, 101
perchlorate, 85
performance, 7, 47, 50, 52, 58, 67, 81, 101, 107, 109
permit, 12, 14, 27
planned action, 64, 109
plants, vii, 4, 49
plastics, 44
polycyclic aromatic hydrocarbon, 88
price caps, 47, 67, 68
private party, 46, 55, 80, 111
privatization, 70
procurement, 67
project, 63, 68, 69, 79, 81
properties, 11, 49, 54, 55, 70, 72, 77, 106, 108

Index

public health, 10, 15, 34, 43, 50, 51, 54, 72, 73, 103, 108, 110, 111
public interest, 81
public resources, 72

R

receptors, 15, 104
recommendations, iv, 37, 46, 63, 79
recurrence, 71
Registry, 20, 43, 47, 50, 81
regulatory oversight, 6, 28
regulatory requirements, 8, 22
reliability, 35
remediation, 3, 9, 32, 58, 66, 69, 79, 101
repair, vii, 4, 49
requirements, 9, 10, 11, 12, 18, 22, 29, 30, 31, 42, 43, 47, 51, 52, 58, 79, 80, 89, 91, 101, 106, 107
resolution, 23, 34, 43, 53, 100, 107
resources, 12, 64, 65, 67, 109
respect, 42, 74, 107, 109
restoration programs, 38
risk assessment, 68, 72
rubber products, 44

S

screening, 41, 52, 75
Secretary of Defense, 24, 42, 54, 79, 101, 111
sediment, 15, 30, 68, 72, 90, 104
sediments, 72
Senate, 48, 49
sequencing, 67
solid waste, 56
solvents, 30, 31, 83, 84, 85, 86
stakeholders, 68, 79, 100
state laws, 107
state oversight, 27, 29, 31
statute, 2, 6, 12, 22, 42, 54, 61, 74, 109
statutes, viii, 10, 43, 46
stigma, 3, 9, 29, 30, 31
storage, vii, 4, 12, 14, 29, 31, 49, 51, 53, 86
survey, 70, 73

T

tanks, 29, 31, 58
testing, vii, 4, 17, 30, 49, 52, 90, 105
threats, 28, 44
time frame, 21, 25, 34, 49, 57, 64, 66, 68, 78, 101, 102, 109
Title I, 43
Title II, 43
toxicity, 16, 104
tracks, 44, 61, 62
training, 31, 43, 89, 90
transparency, 34, 35, 46, 78, 79, 80, 100
transport, 5, 15, 103
transportation, 12, 51

U

U.S. Geological Survey, 31
uniform, 46, 77, 79, 100
universe, 28, 35

V

vessels, 42
vinyl chloride, 90

W

waste, vii, 2, 3, 4, 5, 8, 9, 10, 12, 14, 15, 16, 19, 23, 26, 28, 30, 31, 35, 37, 42, 43, 49, 50, 51, 52, 53, 56, 58, 71, 72, 76, 85, 86, 103, 104, 105, 108
waste disposal, 58
waste management, 28, 43
waste treatment, 12, 14
weapons, vii, 4, 49, 89
welfare, 51
wells, 31, 68, 86

Z

Zone 1, 87